W. Larcher

Physiological Plant Ecology

Translated by M. A. Biederman-Thorson

With 152 Figures

Springer-Verlag
Berlin Heidelberg New York 1975

Professor Dr. WALTER LARCHER
Institut für Allgemeine Botanik der Universität Innsbruck
A-6020 Innsbruck, Sternwartestr. 15

MARGUERITE A. BIEDERMAN-THORSON, Ph.D.
The Old Marlborough Arms, Combe, Oxford, England

Translated and revised from the German edition "Walter Larcher, Ökologie der Pflanzen",
first published 1973 by Eugen Ulmer, Stuttgart. © 1973 by Eugen Ulmer

ISBN 3-540-07336-1 Springer-Verlag Berlin Heidelberg New York
ISBN 0-387-07336-1 Springer-Verlag New York Heidelberg Berlin

Library of Congress Cataloging in Publication Data. Larcher, Walter, 1929–.
Physiological plant ecology. Translation of Ökologie der Pflanzen. Bibliography: p.
Includes index. 1. Botany–Ecology. I. Title. QK901.L3513.581.5. 75-16488.

Typesetting and printing: Carl Ritter & Co., Wiesbaden. Bookbinding: Brühlsche Uni-
versitätsdruckerei, Gießen.

2131/3130-54321

Preface

Ecology is the science of the relationships between living organisms and their environment. It is concerned with the web of interactions involved in the circulation of matter and the flow of energy that makes possible life on earth, and with the adaptations of organisms to the conditions under which they survive. Given the multitude of diverse organisms, the plant ecologist focuses upon the plants, investigating the influence of environmental factors on the character of the vegetation and the behavior of the individual plant species.

Plant ecophysiology, a discipline within plant ecology, is concerned fundamentally with the physiology of plants as it is modified by fluctuating external influences. The aim of this book is to convey the conceptual framework upon which this discipline is based, to offer insights into the basic mechanisms and interactions within the system "plant and environment", and to present examples of current problems in this rapidly developing area. Among the topics discussed are the vital processes of plants, their metabolism and energy transformations as they are affected by environmental factors, and the ability of these organisms to adapt to such factors. It is assumed that the reader has a background in the fundamentals of plant physiology; the physiological bases of the phenomena of interest will be mentioned only to the extent necessary for an understanding of the ecological relationships. A real understanding of plant ecology requires familiarity with methodological problems and their solution; the texts by Šestak et al. (1971) on measurement of productivity parameters and Slavik (1974) on methods in the area of water relations are recommended.

Ecology is very much a modern field, but by no means a recent innovation. I have tried to portray this rich historical background in the choice of illustrations and tabular material; the results presented reflect the broadness of vision, the struggles and the successes of the pioneering experimental ecologists in the first half of this century, as well as the advances in knowledge made most recently.

For this English edition, the original German text has been revised, and in certain places expanded and corrected. Where possible, the German abbreviations have been replaced by those predominant in the English literature. Figures 3, 8 and 89 (formerly Fig. 90) have been extended, and three new figures have been added (Figs. 86, 145 and 152).

My first thanks are due to Dr. K. F. Springer; his publication of this English edition has made the textbook accessible to a wider circle of readers. I am grateful to the publisher of the original German edition, Roland Ulmer, for his cooperation. In particular, I thank Dr. Marguerite Biederman-Thorson for her thoughtful and sympathetic translation into English of the German text.

Above all, however, I should like to express my thanks to the pioneers of experimental ecology—Arthur Pisek, Otto Stocker, Heinrich Walter, and the late Bruno Huber. They inspired my enthusiasm for this difficult, but so attractive, field, and allowed me to benefit from their experience.

Innsbruck, September 1975   W. LARCHER

Contents

Abbreviations, Symbols and Conversion Factors

A	Area	ε_P	Efficiency of radiant energy conversion in terms of productivity of the vegetation
Acc	Acceptor molecule		
ADP	Adenosine diphosphate		
ATP	Adenosine triphosphate	ε_F	Efficiency of radiant energy conversion in terms of photosynthesis
B	Plant biomass (also called phytomass, the mass of a stand of plants)		
		E_p	Evaporative power of the air; potential evaporation
ΔB	Change in biomass (positive for a growing stand)	erg	Unit of energy or work (1 erg = 1 dyn · cm)
bar	Unit of pressure (1 bar = 1 megadyne per cm^2)	Φ	Flux, mass flow
		F	Photosynthesis
C	Concentration	F_g	Rate of gross photosynthesis
C_a	Concentration of CO_2 and H_2O in the air outside a leaf	F_n	Rate of net photosynthesis
		g	Gram; unit of mass
C_i	Concentration of CO_2 and H_2O in the intercellular system of a leaf	G	Grazing (loss of dry matter to consumers)
		GAP	Glyceraldehyde-3-phosphate
cal	Calorie, a unit of energy (1 cal = 4.1868 joule = 4.1868 · 10^7 erg)	h	Hour; unit of time
		ha	Hectare; unit of area (1 ha = 10^4 m^2)
D	Molecular diffusion coefficient	$h\nu$	The energy of a light quantum
d	day as a unit of time		
d	diameter	I	Irradiance; the radiation flux at a given level within a stand of plants or body of water
DL$_{50}$	Drought lethality (degree of dryness causing 50% injury)		
		I_0	Maximum radiation flux; that incident upon a stand of plants or body of water
DM	Dry matter		
dm^2	Unit of area; for leaves, it refers to one (projected) surface		
		I_a	Long-wavelength radiation from the atmosphere
dm$_2^2$	Unit of leaf area referring to the entire surface (upper and lower)	I_{abs}	Absorbed radiation
		I_d	Direct solar radiation
E	Transpiration	I_g	Long-wavelength thermal radiation from ground and plants
E_c	Cuticular transpiration		
E_s	Stomatal transpiration		

P_g	Gross productivity	r_i	Diffusion resistances in the intercellular system
P_i	Inorganic phosphate		
P_n	Net productivity	r_p	Diffusion resistances in protoplasm
π	Osmotic pressure		
PEP	Phosphoenol pyruvate	r_s	Stomatal diffusion resistance
%	Percent (parts per hundred)	r_w	Diffusion resistances in the cell wall
PGA	3-phosphoglyceric acid		
pH	Negative logarithm of the hydrogen ion concentration	r_x	Carboxylation (excitation) resistance
PhAR	Photosynthetically active radiation (400–700 nm)	R	Gas constant ($R = 8.3$ J $\cdot$ deg$^{-1} \cdot$ mol^{-1})
ppm	Parts per million	R	Respiration
Pr	Precipitation (total falling on a stand of plants)	R_d	Dark respiration
		R_l	Respiration in the light
Pr_n	Precipitation reaching the ground beneath a plant canopy	RH	Relative humidity
		RuDP	Ribulose-1,5-diphosphate
		RuP	Ribulose-5-phosphate
Ψ	Water potential	s	Second; unit of time
PWP	Permanent wilting percentage	t	Time (point in time or duration)
Py	Pyruvate	t	Ton (Metric; 1 t $= 10^3$ kg)
Q	Energy flow	T	Temperature (all temperature data in °C)
Q_E	Energy conversion associated with evaporation and condensation	T_a	Air temperature
		T_l	Leaf temperature
Q_H	Energy conversion associated with convection	τ	Matric pressure or potential
		TL$_{50}$	Temperature-stress lethality (the temperature at which 50% of plants are killed by heat or cold)
Q_I	Energy conversion associated with radiation from the sun and reradiation		
Q_M	Energy conversion associated with metabolism	torr	Unit of pressure (1 torr $= 1.33 \cdot 10^{-3}$ bar $\doteq$ a 1-mm column of Hg)
Q_P	Energy conversion in plant communities		
Q_{Soil}	Energy conversion in the soil	UV	Ultraviolet radiation (< 400 nm)
Q_{10}	Temperature coefficient of biochemical and physiological process	W	Watt; unit of power (1 W $= 1$ J $\cdot$ s^{-1})
		W	Weight
r	Transport or diffusion resistance	W_{abs}	Quantity of water absorbed
		W_{act}	Actual water content (when sample is taken)
r_a	Boundary (air) layer resistance	W_{av}	Available water

W_d	Dry weight	*WSD*	Water saturation deficit
W_{FC}	Water content of soil at field capacity	yr	Year
		z	Relative height or depth
W_{PWP}	Water content of soil at permanent wilting percentage	$\varnothing$	Diameter
		$\doteq$	Approximately equal to
W_f	Fresh weight	$>$	Larger than
W_s	Water content in saturated state	$<$	Smaller than

Equivalents

Radiation

$$1 \text{ cal} \cdot \text{cm}^{-2} \cdot \text{min}^{-1} = 6.98 \cdot 10^5 \text{ erg} \cdot \text{cm}^{-2} \cdot \text{s}^{-1} = 6.98 \cdot 10^{-2} \text{ W} \cdot \text{cm}^{-2}$$
$$1 \text{ erg} \cdot \text{cm}^{-2} \cdot \text{s}^{-1} = 1.43 \cdot 10^{-6} \text{ cal} \cdot \text{cm}^{-2} \cdot \text{min}^{-1} = 10^{-7} \text{ W} \cdot \text{cm}^{-2}$$
$$1 \text{ W} \cdot \text{cm}^{-2} = 10^7 \text{ erg} \cdot \text{cm}^{-2} \cdot \text{s}^{-1} = 14.3 \text{ cal} \cdot \text{cm}^{-2} \cdot \text{min}^{-1}$$

$$1 \text{ cal} \cdot \text{cm}^{-2} \cdot \text{min}^{-1} \doteq 60 \text{ kLx}$$
$$100 \text{ kLx} \doteq 1.5 \text{ cal} \cdot \text{cm}^{-2} \cdot \text{min}^{-1}$$

Pressure

$$1 \text{ bar} = 10^6 \text{ dyne} \cdot \text{cm}^{-2} = 1.019 \text{ atm (technical)}$$
$$1 \text{ atm (technical)} = 1 \text{ kg-weight} \cdot \text{cm}^{-2} \doteq 0.981 \text{ bar}$$

Energy Consumption in the Evaporation of Water

$$\text{Heat of vaporization at } 0° \text{ C} = 597 \text{ cal} \cdot \text{g}^{-1} \text{ H}_2\text{O}$$
$$\text{at } 10° \text{ C} = 592 \text{ cal} \cdot \text{g}^{-1}$$
$$\text{at } 20° \text{ C} = 586 \text{ cal} \cdot \text{g}^{-1}$$
$$\text{at } 30° \text{ C} = 580 \text{ cal} \cdot \text{g}^{-1}$$

Phytomass

$$1 \text{ g DM} \cdot \text{m}^{-2} = 10^{-2} \text{ t} \cdot \text{ha}^{-1}$$
$$1 \text{ t} \cdot \text{ha}^{-1} = 100 \text{ g} \cdot \text{m}^{-2}$$
$$1 \text{ g org. DM} \doteq 0.45 \text{ g C} \doteq 1.5 \text{ g CO}_2$$
$$1 \text{ g C} \doteq 2.2 \text{ g org. DM} \doteq 2.7 \text{ g CO}_2$$
$$1 \text{ g CO}_2 \doteq 0.67 \text{ g org. DM} \doteq 0.37 \text{ g C}$$

Further aids to conversion can be found in the *Manuals of Methods* by Šestak *et al.* (1971) and Slavik (1974).

The Environment of Plants

Plants have colonized nearly all regions of the earth, including the oceans and inland waters; on land they can be found even in such inhospitable places as deserts and fields of ice. Far back in geological time, when the first land plants were evolving, they encountered a world of water, air and stone. That is, their environment consisted of the hydrosphere, atmosphere and lithosphere. Later, as the cover of vegetation gradually closed, and with the assistance of microorganisms and animals, there developed the most important substrate of plants: the soil—the pedosphere.

The Hydrosphere

The hydrosphere comprises the oceans of the world, which cover an impressive 71% of the earth's surface, as well as the inland waters and the groundwater. Great differences exist in the chemical compositions of these bodies of water (Fig. 1). Sea water, rich in Na^+, Mg^{2+}, Cl^- and SO_4^2 and with an average salt content of 35 g · l^{-1}, differs fundamentally from fresh water, which usually contains more Ca^{2+} and HCO_3^-; but there are local differences as well, depending on the nature of the inflowing waters and the degree of mixing. Moreover, currents have an effect upon temperature gradients. Where there are no currents, the strong absorption of radiation in the upper levels of the water leads to a characteristic layering with respect to temperature and density; this has a marked influence upon nutrition, productivity and distribution of aquatic organisms (see p. 19).

The Atmosphere

The air enveloping the earth provides plants with carbon dioxide and oxygen. It also mediates the balance of water through the processes of rain, condensation and "evapotranspiration". Continual movement of the air ensures that its composition remains fairly constant—79% nitrogen (by volume), 21% oxygen and 0.03% carbon dioxide, water vapor and noble gases (Fig. 1). In addition the air contains gaseous, liquid and solid impurities; these are primarily sulfur dioxide, unstable nitrogen compounds, halogen compounds, dust and soot.

The part of the atmosphere with which plants come into contact is the troposphere, the weather zone of the earth's envelope of air. The nature of this zone varies over short distances and is characterized in several ways: (1) by the *weather* (short-term events such as showers, thunderstorms and gusts of wind), (2) by meteorological events of intermediate duration such as periods of rain or frost and (3) by the *climate* (the average state and ordinary long-term fluctuations in meteorological factors at a given place). Depending on the terrain and on the density, height and type of vegeta-

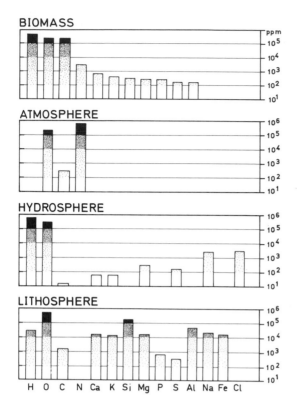

BIOMASS

ATMOSPHERE

HYDROSPHERE

LITHOSPHERE

H O C N Ca K Si Mg P S Al Na Fe Cl

Fig. 1. Composition of the biomass, atmosphere, hydrosphere and lithosphere, in terms of the relative numbers of atoms (atoms per million atoms, not the proportion by weight) of the various chemical elements. The composition of living organisms is clearly distinct from that of the three components of their environment; they select from the available elements, according to their needs. The scale of the ordinate is logarithmic. For example, in the biomass H, O, C and N are present in the greatest proportions: $4.98 \cdot 10^5$ atoms per million (i. e., about 50% of all atoms) are hydrogen atoms; oxygen and carbon atoms each comprise $24.9 \cdot 10^5$ atoms per million (about 25%), and $2.7 \cdot 10^3$ (about 0.3%) are nitrogen atoms. (After Deevey, 1970)

tion, individual climatic regions of different sizes are formed. Within the large-scale "macroclimate" measured by the network of meteorological stations, one may distinguish "microclimates" that prevail in specific places such as certain slopes, narrow valleys and stands of vegetation; an "interface" climate—in the layer of air near the ground and the surface of leaves—may also be distinguished. Thus the parts of plants above ground are exposed to variability, in space and time, with respect to radiation, temperature, humidity, precipitation and air motion; any of these can from time to time represent a threat to the organism.

The Lithosphere and the Soil

The **earth's crust** is the inexhaustible storehouse for the variety of chemical elements of which organisms are composed (Fig. 1). The lithosphere exchanges matter with the hydrosphere, and also affects the composition of the atmosphere through volcanic action and the products of radioactive decay. Primarily, however, it is the basic material for the formation of the soil.

Soil is more than just superficially loosened lithosphere. It is the product of the transformation and mingling of mineral and organic substances. Soils are produced with the assistance of organisms and under the influence of environmental conditions. They are subject to continual change: soils grow, mature, and can age and perish. Physical and chemical weathering continuously frees mineral substances from the rocky substrate, and there is an unceasing decay of plant remains and dead organisms. These decay products, together with the excrement of soil animals, gradually turn into humus, which forms complexes with the mineral products of weathering. In natural soils a profile is established of more or less horizontal layers ("horizons"): between the strata of litter and humus, and the stratum where weathering of the parent rock occurs, there are transitional zones with varying proportions of humus. The types and thicknesses of the horizons in such a profile are characteristic of a given type of soil and reflect the influence of climate, plant cover, soil organisms, underlying rock and the activity of man. Pedology is the science of soil formation and composition, and of the classification of soil types. Knowledge of the fundamentals of this field is an absolute prerequisite to understanding plant ecology.

The solid particles in the ground stick together to form aggregates, leaving small open spaces. Together these form a system of pores penetrating the entire soil, filled partly with air and partly with water. Thus the soil is a three-phase system in which lithosphere (the solid phase), hydrosphere (fluids) and atmosphere (gases) are intermingled. It has an enormous capacity for uptake and storage, and is particularly suited for the buffering of physical and chemical influences. Below the top centimeters of soil the prevailing climate is more stable than that of the atmosphere; radiation is essentially unable to penetrate, there are no sharp gradients of temperature, and the processes of exchange are slow, occurring by diffusion. Therefore the soil is the most suitable habitat for many organisms. The roots—in many respects the most vulnerable organ system of the higher plants—are entirely adapted to life in the soil. A landscape without soil is a life-repelling "lunar" landscape. Only a few remarkable plants such as aerial algae, lichens and mosses can actually thrive on bare stone or sand.

The Biosphere

Atmosphere, hydrosphere and lithosphere existed before there was life on earth, but it is of course only through the appearance of living organisms that they became significant as an "environment". "Environment", as defined by A. F. Thienemann, is "the totality of external conditions affecting a living organism or a community (biocenosis) of organisms in its habitat (its biotope)". In this strict sense, only living beings have an environment. It comprises not only the influences exerted by the abiotic surroundings, but also those due to the other organisms present.

The part of the earth which supports life, called the biosphere, is the narrow band, about 100 m thick, above the earth and below the surface of the ocean, which is ordinarily inhabited by organisms. It is true that birds can fly as high as 2,000 m, and that there are bacteria on the floor of 10,000-m-deep marine trenches, but abundant life is limited to a much more restricted region near the surface of the earth. Trees stand no more than 70—100 m tall and sink their roots but a few tens of meters into the earth; in water the layer penetrated by light (and hence densely populated) ordinarily extends to a depth of 30 m, and at most to 100 m.

Among living organisms, the plants are of prime significance. They are capable of capturing and storing by photosynthesis the energy from outer space—that of sunlight—and in terms of mass they far exceed all other organisms; about 99% of the total mass of living beings (the biomass) on earth is accounted for by the plants (the phytomass). Because of this enormous mass, the plant cover is a stabilizing factor in the cycling of matter and has a crucial effect upon the climate.

The Ecosystem: Interplay of Biological and Environmental Factors

Communities of organisms and their abiotic surroundings interact in nature by many kinds of reciprocal relationships, both structural and functional. Circumscribed, more or less uniform sections of the biosphere delimit biogeocenoses, or ecosystems. The expression "biogeocenosis" was coined by V.N. Sukachev, and the term "ecosystem" is attributable to R. Woltereck and A.G. Tansley. H. Ellenberg defines the ecosystem as "a unitary system of interactions involving living organisms and their inorganic environment which is, to a certain degree, capable of self-regulation". In short,

$$\text{Ecosystem} = \text{community of organisms} + \text{environmental conditions} \qquad (1)$$

Each ecosystem has a certain spatial extent; together ecosystems form a diverse mosaic in the biosphere. A forest is an ecosystem, as is a meadow, a lake or an ocean. The principles by which all these ecosystems operate hold equally well for the biosphere as a whole, and as well for natural ecosystems as for artificial systems such as aquariums and self-contained manned spacecraft. Fig. 2 represents the typical structure of an ecosystem and the most important interactions both within it and between it and the external world.

The Components of an Ecosystem

Every independently functional ecosystem is composed of at least two biological components: the producers and the decomposers. Between these there may exist a whole chain of consumers.

Primary Producers are the autotrophic organisms, which effect the incorporation of inorganic elements in organic compounds and thus raise them to a higher energy level. The green plants and some bacteria utilize sunlight to form carbohydrates from carbon dioxide and water (photosynthesis), and these become the basis for further syntheses. Various microorganisms accomplish the same thing by using the energy freed by exergonic inorganic reactions (chemosynthesis).

Consumers (or phagotrophs) are the heterotrophic organisms that feed directly or indirectly upon the organic substances synthesized by the primary producers. The principal consumers are the herbivores and the plant parasites. The herbivores serve in turn as food for the carnivores, and both are attacked by animal parasites.

Decomposers (or saprotrophs) are those organisms that finally reduce plant and animal refuse to the level of its basic inorganic components. This group includes

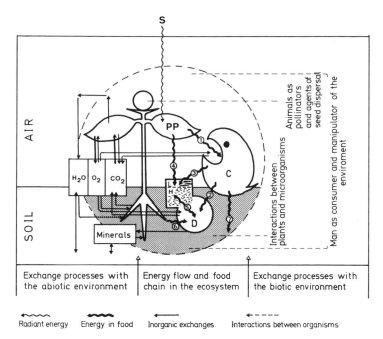

Fig. 2. The flow of energy and matter, and the cycling and exchange of substances, in a terrestrial ecosystem. *Components of the ecosystem: PP* primary producers; *C* consumers; *D* decomposers; *L* deposits of detritus (plant litter, bodies of animals); *H* humus. *Energy flow and the food chain; S* radiation from the sun; *1* consumption of plants as fodder and by parasites; *2* organic excretions from animals and microorganisms; *3* detritus consisting of animal cadavers and dead microorganisms; *4* detritus derived from the primary producers; *5* decomposition (humification and mineralization) of detritus; *6* organic excretions from the plants; *7* loss of organic waste from the ecosystem. *Inorganic transport of substances:* CO_2 to primary producers (photosynthesis) and from processes of breakdown in primary producers, consumers and decomposers (soil respiration), O_2 from primary producers (photosynthesis) to oxygen-consuming catabolic processes; H_2O from evaporative surfaces (from the ground and through organisms) into the atmosphere, as precipitation from the atmosphere into the soil, from the soil through consumption by organisms and loss by drainage; mineral substances from the soil into primary producers and back again *via* the food chain and the activity of decomposers (mineralization)

bacteria and some soil animals. The decomposers, like the herbivores and other consumers, can serve as food for other organisms. In such cases, they take on the role of *secondary producers*. Thus a single individual, depending on its position in the food chain, can play the role of secondary producer, consumer or decomposer.

Food Chains and Energy Flow in an Ecosystem

A "food chain" is defined as the sequence of stages through which stored energy in the form of food is passed from the primary producers to a number of organisms. As

5

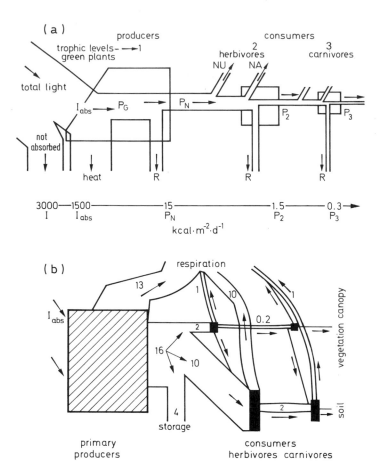

Fig. 3. (a) Simplified energy-flow diagram depicting three trophic levels (boxes numbered *1, 2, 3*) in a linear food chain. Standard notations for successive energy flows are as follows: I = total energy input; I_{abs} = light absorbed by plant cover; P_G = gross primary productivity; P_N = net primary productivity; P_2, P_3 = secondary (consumer) productivity; NU = energy not used (stored or exported); NA = energy not assimilated by consumers (egested); R = respiration. Bottom line in the diagram shows the order of magnitude of the energy losses expected at major transfer points, starting with a solar input of 3000 kcal per square meter per day. (b) Two-channel energy flow diagram that separates a grazing food chain (vegetation canopy) from a detritus food chain in a hypothetical forest. Estimates in kcal $\cdot$ m^{-2} $\cdot$ day^{-1}. (After Odum, 1971)

the organisms at each stage are eaten by those of the next, a flow of energy through the ecosystem is maintained, beginning with the capture and conversion of radiant energy from the sun and ending with the complete decomposition of the organic compounds; at each consumer level, some energy is lost. An example is shown in Fig. 3.

The Cycling of Matter

The net transfer of energy in a chain always proceeds in one direction—from the sun through the primary producers to the decomposers. Matter, on the other hand, is cycled. Each cycling of bioelements takes place at three distinct levels: in the plant, in the ecosystem and in the biosphere. The individual organism takes up substances, incorporates them in its body or obtains energy from them by metabolism, and excretes them. At the second level, the movement of materials through the ecosystem begins with conversion by the metabolism of the primary producers and leads back to the starting point through intermediate consumers and the decomposers—i.e., through the food chain. The food cycle in the ecosystem is associated with geochemical transpositions of matter which extend beyond the ecosystem itself; these exchanges link the metabolism of the organisms with the transformations of inorganic chemicals normally occurring in the biosphere. Because of this complex biogeochemical exchange of matter, each ecosystem itself is actually an open system.

There are two fundamental types of biogeochemical cycles of matter. In the gaseous type, in which the participants are carbon, oxygen and water, the atmosphere and the hydrosphere constitute the most important reservoirs, and large-scale cycling takes place relatively rapidly. In contrast, in the earthbound ("sedimentary") cycles, sulfur, phosphorus and the other mineral bioelements are transported in aqueous solution—the stores available in the soil and the earth's crust are much less mobile than those in the gaseous cycle. The nitrogen cycle occupies an intermediate position. Plants draw nitrogen from the soil in the form of ions, but the chief store of the element is in the form of atmospheric gas.

Autoregulation

Ecosystems possess intrinsic mechanisms for self-regulation, based both on the pronounced ability of individual organisms, populations and communities to adapt, and on the properties of the closed sequence of producers, consumers and decomposers in the biological nutrient cycle. If the supply of energy is sufficient and the recycling of matter undisturbed, then once a state of equilibrium has been reached with respect to the interrelationships of species and the exchange of materials with the environment, the state is maintained without notable change. Such *stable ecosystems* consist of a large biomass which does not further increase, since production balances consumption in the nutrient cycle over a year. E.P. Odum calls these ecosystems *protective*. The model of a protective ecosystem is the tropical rain forest—the plants produce a great deal of organic substance but a large part of it enters the cycle and is remineralized. This point will be discussed in more detail in the section on the cycling of carbon and various minerals. When land is newly colonized, or after drastic damage such as that caused by fire in forest or heath, a state of equilibrium in the ecosystem must be gradually established. Young ecosystems are at first susceptible to disturbance—they are *unstable*. Their relatively small biomass does not fully occupy the available space, but their rate of growth is high; they are *productive*. As they mature, their rate of growth is regulated, their productivity approaches a state of equilibrium, and the initially productive ecosystem eventually becomes a protective ecosystem. Autoregulation ensures that the development of mass and the increase in number of members of the system remain subordinate to the functional whole—the population does not increase without bound.

The readiness to adapt and the autoregulatory capacity of a biological community can be overburdened by extraordinary environmental events. Such "stresses" may be transitory though severe—e.g. floods, avalanches, frost, fire, drought and outbreaks of parasites—but they can also be long-term disturbances involving air and water pollution or unrestrained exploitation by man. Depending on their species composition, structure and the nature of their equilibrium state, ecosystems are differentially "resilient" in the face of such factors. In H. Ellenberg's terminology, the "resilience" (*Belastbarkeit*) of an ecosystem is greater the less susceptible it is to disturbance and the more capable it is of recovering rapidly and completely when the disturbance is removed.

Biotic Interactions

In addition to the food chain, there are other kinds of interdependence and interaction relationships among the organisms in an ecosystem that may either benefit or hamper the individual. Examples of *beneficial* biotic influences include ecological relationships between plants and animals that promote pollination and distribution; moreover, plants can live in symbiosis with microorganisms (cf. p. 94), and full-grown plants can shield seedlings from intense radiation, overheating, or cooling (cf. p. 217). *Inhibitory* influences include the competition among plants for light, water and nutrients, and the repulsion and suppression of organisms by secretions (allelopathy between adjacent plants, antibiosis between microorganisms), as well as parasitism in all its manifestations (for example, hemiparasites such as mistletoe interfere with the water balance of the host plant).

Man and the Biosphere

Like any other organism, man is an environment-dependent member of the living community and an environmental factor with respect to other members of the ecosystem in which he happens to participate. But he is more than that. To the extent that his behavior is determined by intelligence, his cultural development has raised him above the narrow confines of limited biocenoses and made him a citizen of the world. As a conscious and purposeful shaper of nature, he intervenes with incomparable effectiveness in natural processes and creates new habitats to his own liking. But he can also misuse these unique abilities, disturbing or destroying every habitat on the planet—especially if his knowledge fails to comprehend environmental relationships and if his will is not tempered by a sense of responsibility.

The Sun's Radiation as a Source of Energy

All life on earth is supported by the stream of energy radiated by the sun and flowing into the biosphere. Even the relatively small amount of radiant energy bound in the form of latent chemical energy by the photosynthesis of plants suffices to maintain the biomass and the vital processes of all members of the food chain. By far the larger fraction of radiation absorbed is transformed immediately into heat; part of this fraction is used in the evaporation of water and the rest produces an increase in the temperature on the earth's surface. Radiation is thus the source of energy underlying the distribution of heat, water and organic substances. It creates the prerequisites for an environment adequate to sustain life.

Radiation within the Atmosphere

The Spectral Composition of Sunlight

The biosphere receives solar radiation in the range of wavelengths between 290 nm and about 3,000 nm. Radiation at shorter wavelengths is absorbed in the upper atmosphere by ozone and atmospheric oxygen, and the long-wavelength cutoff is caused by the air's content of water vapor and carbon dioxide. About 40—45% of the energy incident from the sun consists of wavelengths between 380 and 720 nm. This is the region of the spectrum that we perceive as visible light. The chloroplast pigments absorb radiation between about 380 and 740 nm (cf. Fig. 6). Ultraviolet radiation (UV) bounds the photosynthetically active radiation (PhAR; usually considered as lying between 400 and 700 nm) on the short-wavelength side, and infrared (IR), on the long-wavelength side.

Attenuation of Radiation by the Atmosphere

At the outer limits of the earth's atmosphere the intensity of radiation is nearly $2 \text{ cal} \cdot \text{cm}^{-2} \cdot \text{min}^{-1}$ (more precisely, the "solar constant" is $1.94 \text{ cal} \cdot \text{cm}^{-2} \cdot \text{min}^{-1}$, or $1.39 \text{ kW} \cdot \text{m}^{-2}$). Of this, however, only an average of 47% reaches the earth's surface. Of the incident sunlight, 25% is reflected by clouds and 9% is scattered by atmospheric particles and returned to outer space. A further 10% is absorbed by clouds, and 9% by water vapor. The radiation eventually reaching the ground or the plant cover is composed of direct sunlight (averaging 24% of the incident radiation), diffuse radiation from clouds (cloudlight, 17%), and diffuse radiation from the sky (skylight, 6%). These values are long-term average values for the northern hemisphere. At sea level a horizontal surface at intermediate latitudes at noon receives total (direct plus diffuse) radiation amounting to as much as $1.3 \text{ cal} \cdot \text{cm}^{-2} \cdot \text{min}^{-1}$. Depending on the latitude of a site, its altitude above sea level, the nature of the terrain and the frequency of clouds, there are large regional and local differences in

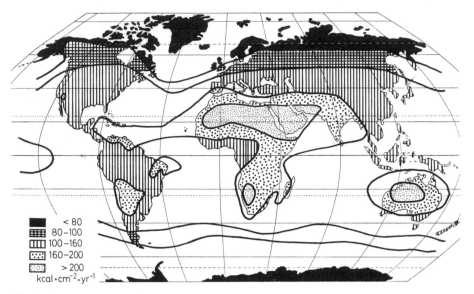

Fig. 4. Annual average of solar radiation (300—2200 nm) reaching the surface of the earth. (From Geiger, 1965)

the supply of radiation (Fig. 4). Thus, the high-pressure regions in the tropics, where clouds are few, receive a greater than average quantity of solar radiation; rather than 47%, an average of 70% of the incident radiation penetrates the clear envelope of air above the dry regions. Moreover, at greater altitudes, owing to the shorter optical path of the rays and the lesser degree of air turbidity, more radiation reaches the ground than in lower-lying places.

Uptake of Radiation by Plants

The Leaf as a Radiation Receiver

Some of the radiation striking a leaf is reflected at the surface, some is absorbed so as to be physiologically effective, and the remainder is transmitted through the leaf. The degree of reflection, absorption and transmission in leaves depends on the wavelength of the radiation (Fig. 5).

Reflection

In the infrared region, leaves reflect 70% of radiation incident perpendicularly, whereas in the PhAR range an average of only 6—12% is reflected. Green light is more strongly reflected (10—20%), orange and red light less so (3—10%). Little ultraviolet radiation is reflected; as a rule, in the UV range, leaves reflect no more than 3%. Certain flowers, however, display marked UV reflection that can be detected by insects and serves as an attractive target.

10

Fig. 5. Relative reflection, transmission and absorption of a poplar leaf (*Populus deltoides*) as a function of the wavelength of the incident radiation. (After Gates, 1965)

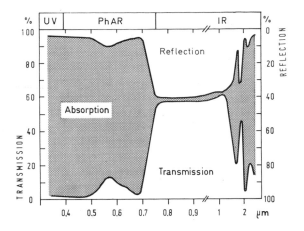

The capacity to reflect light depends upon the nature of the leaf surface; for example, a dense covering of hairs can increase reflection by a factor of two or three.

Absorption

Radiation penetrating a leaf is to a great extent absorbed. Most of the UV is retained by the epidermis; only 2—5% of the UV radiation enters the deeper levels of the leaf. Thus the epidermis is an effective UV filter, protecting the parenchyma in which photosynthesis occurs. PhAR absorption is determined by the chloroplast pigments. Correspondingly, the spectral absorption curves of leaves show maxima wherever the absorption maxima of chlorophylls and carotenoids occur. About 70% of the photosynthetically utilizable radiation entering the mesophyll is absorbed by the chloroplasts. In its passage through the leaf, radiation is progressively attenuated, so that the amount captured by successive cell layers falls off approximately exponentially. Not much infrared is absorbed in the region up to 2 μm, but in the range of long-wavelength heat radiation above 7 μm it is almost completely (97%) absorbed. Accordingly, the plant behaves like a black body with respect to heat radiation.

Transmission

Transmission by leaves depends on their structure and thickness. Mesomorphic leaves pass 10—20% of the sun's radiation, very thin leaves pass up to 40% and thick, solid leaves may not transmit any at all. Transmission is greatest at the wavelength ranges where reflection is great—that is, in the green and particularly in the near infrared. Radiation filtered through foliage is therefore particularly rich in wavelengths around 500 nm and over 800 nm. Beneath a canopy of leaves a red-green shade prevails, and in the depths of a forest only red shade remains.

The Effects of Radiation on Plant Life

In the plant, radiation acts as the source of energy for photochemical reactions and as a stimulus regulating development. But it can also cause injury. The various effects of radiation result from the capture of quanta, the energies of which are a function of wavelength.

Only absorbed radiation can be effective; therefore each radiation-dependent process is mediated by specific receptors. The radiation receptors for photosynthesis are chlorophyll and accessory plastid pigments, and in algae phycobilins are also involved. Photo-induced processes such as the timing of germination, flowering, leaf abscission, and in some plants the annual alternation between active and resting states, are controlled by the phytochrome system. In addition, this system controls a number of photomorphoses—light-dependent differentiation of tissue and determination of shape. Directional growth in response to light (phototropism) is mediated by blue-light receptors of the flavoprotein or carotenoid type. The absorption spectra of some important plant pigments, together with the radiation available at different levels in the atmosphere and hydrosphere, are shown in Fig. 6. The long-wavelength end of the range of absorption by nucleoproteins is also indicated; this marks the absolute limit for survival of protoplasm. UV below 280 nm would damage the genetic material, but such short-wavelength radiation does not occur in the biosphere.

In their metabolism, form and development, plants adapt themselves in many ways to the prevailing quality and quantity of the locally available radiation. Examples

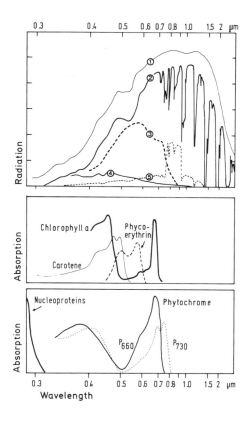

Fig. 6. Spectral distribution of energy in solar radiation outside the earth's atmosphere (*1*), in the direct solar radiation at sea level (*2*), in the radiation 1 m below the water surface in the coastal regions of the ocean (*3*), in the diffuse radiation from the sky (*4*), and in the radiation under a stand of plants (*5*). (After Gessner, 1955; Gates, 1965; R. Schulze, 1970). Below: The spectral absorption by photosynthetic and photomorphogenetic pigments and by nucleoproteins. (After Blinks, 1951; French and Young, 1956; Siegelman and Butler, 1965)

Table 1. A comparison of sun leaves and shade leaves (based on the findings of various authors in *Fagus* and *Solidago*). (From Lorenzen, 1972)

	Sun leaves	Shade leaves
Characteristic structure	smaller blade thicker cuticle smaller surface area, but more stomata	
Cell-sap concentration	+ +	+
Transpiration	+ +	+
Water content	85% of W_f	90% or more
Respiration	+ +	+
Light intensity for CO_2 compensation	high	low
Light intensity for saturation of photosynthesis	high	low
RuDP carboxylase	+ + +	+
Chlorophyll per W_d	+	+ +
Soluble proteins ($\doteq$ enzymes)	+ +	+
Amino acids and raw protein	+	+ + +
Starch	+ +	+
Cellulose	+	+ +
Lignin	+ +	+
Ash	+ +	+
Ca fraction of ash	+ +	+
K fraction of ash	+	+ +
Lipids	+ +	+

include the chromatic adaptation of algae (cf. p. 21) and the differentiation of light and shade leaves (Table 1); in the latter case, secondary effects of radiation (associated primarily with water balance) also play a role.

The Distribution of Radiation in Tree Canopies and in Plant Communities

Brightness Zones in the Crowns of Trees

In vascular plants photosynthesis occurs within a stacked arrangement of leaves which partially overlap and shadow one another. In the crown of a tree, for example, the individual leaves receive unequal amounts of radiation, according to their orientation to the incident light and their position within the mass of leaves. The shape of the crown and the density of the foliage, the predominant factors here, determine a characteristic radiation profile, from the surface of the crown, which is flooded with light, through zones of decreasing intensity to the interior of the crown (Fig. 7).

Around the turn of the century, J. v. Wiesner emphasized the importance of considering the "*relative irradiance*" *(relativer Lichtgenuß)*, expressed as the average percentage of the external light, in order to classify illumination levels in or below a plant stand. Within dense tree crowns—for example, in that of a columnar cy-

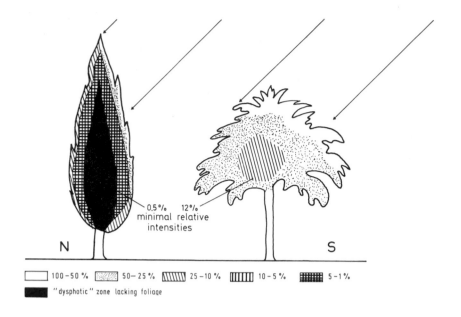

Fig. 7. Attenuation of light in the dense crown of a cypress (left) and the open crown of an olive tree (right), at noon on clear days in July and August. The intensity of illumination at various places in the crown is given as a percentage of the illuminance in the open air. In the dense cypress crown there is a sharp decline in intensity even in the outermost regions, and when the relative intensity becomes less than 0.5% the assimilative shoots turn yellow and dry up. The more open olive crown, with its small, strongly reflecting leaves, is penetrated by diffuse light, so that there are leafy branches even in the darkest part of the crown. (From unpublished data)

press—light can be so attenuated that no new assimilative organs are formed and the old shoots wither. The average light available at the inner boundary of the region of foliage in the crown can be used to infer the particular radiation requirements of the species for photosynthesis and metabolism; that is, the less light needed by the shade leaves of a tree, the more densely packed the crown can be. In open crowns (birch, larch, Scotch pine) this minimal value lies between 10 and 20%, and in dense crowns (spruce, fir, beech), between 1 and 3% of the incident radiation.

Uptake of Radiation in a Plant Community

A stand of plants in which many layers of foliage are superimposed utilizes the incident radiation to the utmost. Owing to repeated reflection and stepwise absorption in the dense zones, very little light penetrates to the ground—only a few percent of the intensity incident upon the stand (Fig. 8). The shade plants comprising the undergrowth are adapted to these low levels of illumination and can thrive even with relative intensities of 5—20%. In general, a relative intensity of 1—2% is considered the lower limit for the existence of vascular plants. Thallophytes can survive in

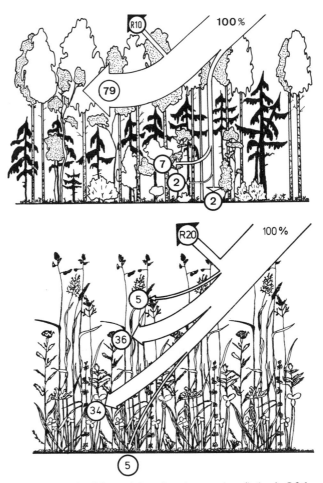

Fig. 8. Attenuation of radiation in a boreal mixed forest (above) and a meadow (below). Of the incident photosynthetically active radiation, 10% is reflected (R) from the upper surface of the forest, and 20% from that of the meadow. Different amounts of the radiation that penetrates the stands are absorbed in each layer, depending on the structure of the stand. In the forest, the greatest absorption of radiation occurs at the crowns of the trees, while in the meadow absorption is greatest in the middle and lower regions, where the leaves are most dense. Only 2 to 5% of the incident PhAR reaches the ground. (After Cernusca, 1975; Kairiukštis, 1967)

locations with even less light, at relative intensities as low as 0.5%, or in the extreme case 0.1% (aerial algae).

Attenuation of Radiation within the Plant Canopy

The fall-off of radiation in a stand of plants depends chiefly on the density of the foliage and the arrangement of the leaves. The foliage density can be expressed

15

quantitatively by the leaf-area index (LAI). This indicates the total surface area of the leaves above a certain area of ground:

$$LAI = \frac{\text{Total leaf area}}{\text{Ground area}} \qquad (2)$$

Ordinarily the units used for leaf area and ground area are identical (m^2), so that LAI is actually a dimensionless measure of the amount of cover. With a LAI of 4, a given area of ground would be covered by four times that area of leaves—arranged in several layers, of course. On its way through the plant canopy, the radiation must pass these successive layers of leaves. In the process, its intensity decreases almost exponentially with increasing amount of cover, in accordance with the *Lambert-Beer Extinction Law*. If the layering of the foliage is taken to be homogeneous, the fall-off of radiation can be computed from M. Monsi and T. Saeki's modification of the extinction equation,

$$I = I_0 \cdot e^{-k \cdot LAI} \qquad (3)$$

where I is the intensity of the radiation at a certain distance from the top of the plant canopy (the irradiance)

I_o is the radiation incident on the top of the canopy

k is the extinction coefficient for this particular plant community

LAI is the total leaf area above the level at which I is estimated, per unit ground area (the cumulative LAI).

The extinction coefficient indicates the degree of attenuation of light within the canopy for a given area index. In grain fields, meadows and clumps of reeds, where the leaves tend to have an upright orientation (more than $^3/_4$ of the leaves are at an angle of more than 45° from the horizontal), the extinction coefficient is less than 0.5, and in the middle of the stand the light intensity is still at least half that of the external light (Figs. 9 and 10: *graminaceous type*). In contrast, for plant communities with broad, horizontal leaves such as fields of clover, tobacco plantations, or stands of tall perennial herbs, the extinction coefficient is greater than 0.7, and at "half height" $^2/_3$ to $^3/_4$ of the incident light has been absorbed (Figs. 9 and 10: *dicotyledon type*).

The relationship between foliage density and light transmission holds for stands of trees as well. Forests with closely packed crowns and dense foliage swallow up so much radiation that near the trunks and on the ground there is very little. In such forests the attenuation of light is similar to, or even more abrupt than, that under dicotyledonous herbs. Woods comprising tree species with sparse crowns (birches, oaks, Scotch pines, eucalyptus) and open stands of trees, on the other hand, attenuate the light as gradually as do grass communities.

Seasonal Alternation in Available Light

During a plant's life cycle the amount of foliage of course varies, so that light distribution in stands of plants changes with the seasons. This is true for deciduous

16

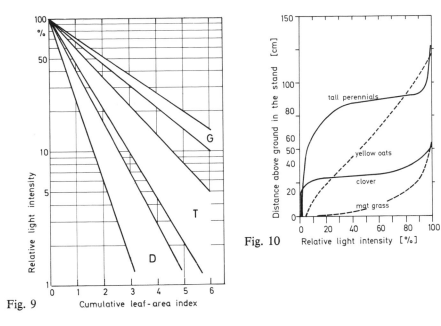

Fig. 9   Cumulative leaf - area index

Fig. 10   Relative light intensity [%]

Fig. 9. The exponential decrease of light intensity in different stands of plants as a function of leaf-area index. The cumulative *LAI* is derived by summation of the index values for the individual horizontal layers of assimilation surface in the stand. In broad-leaved dicotyledonous communities (D), the attenuation of light is considerable even with a low *LAI*, whereas in grass communities (G) attenuation occurs more gradually; stands of trees (T) represent an intermediate position. (After Monsi and Saeki, 1953; Kira et al., 1969)

Fig. 10. The attenuation of light in fields of yellow oats (the consociation Trisetetum trollietosum), low mat grass and yellow oats (Trisetetum nardetosum), tall perennials (Adenostyleto-Cicerbitetum), and clover (of the species *Trifolium subterraneum* and *Lolium rigidum*). In grass communities with relatively upright leaves the intensity falls off smoothly, whereas in dicotyledonous communities with widespread leaves there is a marked decline in intensity even in the upper layers. (After G. and R. Knapp, 1952; Donald, 1961)

forests as well as for communities of herbs, where the full density of foliage and the corresponding attenuation of radiation occur only when the shoots have grown to full size. In winter, 50—70% of the light penetrates the bare deciduous forest to the ground, and while the leaves are developing it is still possible for 20—40% to penetrate; but when the foliage is mature and the crowns of the trees form a closed roof, less than 10% is transmitted (Fig. 11). The ground flora is adapted to this seasonal change in illumination. The life cycles of spring-flowering plants growing under deciduous trees are timed to take advantage of the brief period between the thawing of the snow and the development of foliage by the trees. Young trees, e.g. the seedlings of oak and beech, occasionally lose their leaves much later than do the mature trees and thus can utilize the autumn light. Beneath the crowns of evergreen trees the light intensity remains low throughout the year, and in such forests the undergrowth eventually consists entirely of cryptogams.

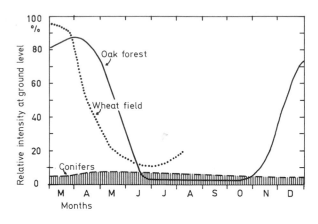

Fig. 11. Seasonal changes in the penetration of light through various stands of plants. Under the shield provided by the crowns of the evergreens in a coniferous forest, the intensity remains low the year round, while in a deciduous forest it depends on the stage of foliage development; in a field of grain it depends on the stage of growth of the plants. (After Nägeli *et al.* as cited by Sauberer and Härtel, 1959)

Distribution and Effects of Radiation in Bodies of Water

The Attenuation of Radiation in Water

In water, radiation is more strongly attenuated than in the atmosphere. Long-wavelength heat radiation is absorbed in the upper few millimeters, and infrared radiation in the uppermost centimeters. UV penetrates the top decimeters or meters. Photosynthetically active radiation (particularly that at wavelengths in the region of 500 nm) reaches greater depths, where blue-green twilight predominates in the ocean and yellow-green, in lakes. The radiation available in bodies of water depends on the following factors:

1. The intensity and nature of illumination above the water level.

2. The amount of reflection and backward scattering of light at or near the water surface; on the average, when the sun is high, a smooth surface reflects 6% of the incident light and a surface with pronounced waves reflects about 10%. When the sun is low, on the other hand, reflection is considerably increased, so that a large fraction of the light fails to enter the water. The consequence is that, under water, the "day" is shorter than on the land.

3. The attenuation as the rays pass through the water; the intensity of radiation decreases exponentially with increasing depth. Radiation is absorbed and scattered by the water itself as well as by dissolved materials, suspended particles of soil and detritus, and plankton. In turbid, flowing waters the light can decrease to 7%—a value comparable to that beneath the crowns of the trees in a spruce forest—at depths of little more than 50 cm. In clear lakes 1% of the incident PhAR reaches depths between 5 and 10 m, so that vascular plants can exist at 5 m and sessile algae, as deep as 20—30 m. The layer of water above the limit for existence of autotrophic plants is called the euphotic zone (Fig. 12). In the open ocean the illuminated euphotic zone is deeper than in lakes; in the Mediterranean near the coast 1% of the radiation penetrates to 60 m, and in the clear water of the oceans the corresponding depth is as great as 140 m.

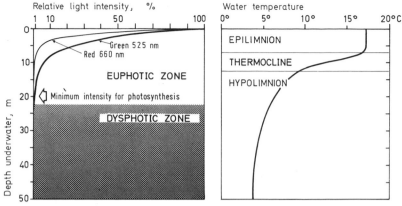

Fig. 12. Decrease in radiation with depth and temperature layering in a typical eutrophic lake of the temperate zone in summer (the Mondsee in Austria). (After Findenegg, 1967, 1969)

Temperature Distribution and Density Layering in Water

Absorption of radiation in the uppermost layers of a body of water warms only these layers. Warm water is less dense than cold, and the currents produced by the wind compensate for density gradients only down to a limited depth; therefore, in the high-radiation season of the year, a very stable, temperature-determined density layering is created in standing waters (Fig. 12). In lakes, sharply demarcated masses of water are produced; a more strongly irradiated (and thus warmer and lighter) superficial layer, the epilimnion, lies upon the deeper, colder, denser water, the hypolimnion. There is a relatively sharp transition between the two, called the thermocline. In the fall the surface water cools off, the thermal layering disappears, and the lake becomes completely mixed. It is thus characteristic of the thermal relationships in a lake that the temperature changes with the seasons in the epilimnion and remains uniformly low in the hypolimnion.

In the ocean there is a permanent thermocline deep under the euphotic layer; an additional, seasonal thermocline becomes established only in temperate latitudes. There, seasonal changes in temperature are also measurable in the top 15—40 m. Near the equator and the poles, the water temperature in the open sea remains constant throughout the year. An additional critical factor with regard to the oceans, of course, is the presence of currents.

Carbon Utilization and Dry Matter Production

Carbon Metabolism in the Cell

Photosynthesis

In photosynthesis, radiant energy is absorbed and transformed into the energy of chemical bonds; for every gram-atomic-weight of carbon taken up, potential energy amounting to 112 kcal is obtained. Photosynthesis involves photochemical processes that occur in the presence of light, purely enzymatic processes not requiring light (the so-called dark reactions), and the processes of diffusion which bring about exchange of carbon dioxide and oxygen between the chloroplasts and the external air. Each of these subprocesses is influenced by internal and external factors and can limit the yield of the overall process. In the following discussion bioenergetic and biochemical aspects of photosynthesis are considered only insofar as they are of ecological importance. It is taken for granted that the reader is familiar with the physiological fundamentals; detailed presentations of these can be found in textbooks of plant physiology, microbiology and biochemistry.

The Photochemical Process

The photochemical process is initiated when the chloroplasts capture photosynthetically utilizable radiation. This capture is accomplished by accessory pigments as well as chlorophyll. Two groups of pigments are involved, and are associated with two different photochemical processes. In pigment system I the light energy is taken up by chlorophyll a and a pigment with an absorption peak at 700 nm (called P700); pigment system II consists of chlorophyll a and accessory pigments most effective at 680 nm. P700 acts as a reaction center for both systems; when excited, it releases electrons that return to the oxidized chlorophyll molecule *via* several redox systems. This cyclic electron transport is coupled with the formation of ATP (*cyclic phosphorylation*). The electrons released by P700 can also be used for the reduction of NADP$^+$. In this case the electrons required for the re-reduction of chlorophyll are provided by the photolysis of water (the Hill reaction). Oxygen is thereby set free, and appears in the overall gas exchange associated with photosynthesis. Pigment system II transfers the electrons obtained from the breakdown of water to P700. ATP is also formed in this *non-cyclic electron transport*. For the transport of one electron, two light quanta (each with energy hv) are required. The course of the light-dependent reactions can be formulated as follows:

Cyclic photophosphorylation:

$$ADP + P_i \xrightarrow[P700]{2\ hv} ATP \tag{4}$$

[handwritten annotations: LIGHT, ELECTRONS, Back t chlorophyll, TRANSFER OF ELECTRONS, through series, redox's, P680 + xhv, H₂O hydrolysis]

20

Non-cyclic electron transport:

$$2H_2O + 2NADP^+ + 2ADP + 2P_i \xrightarrow[\text{P700 + P680}]{8\ h\nu} 2NADPH_2 + 2ATP + O_2 \quad (5)$$

The yield of the photochemical reaction depends upon the amount of energy supplied—that is, upon the *radiation dose*, the product of the intensity of absorbed radiation and the duration of irradiation. It is thus not sufficient to know only the intensity at a given time; one must also determine for how long it is effective.

Photosynthetic Efficiency and Pigment Concentration

The influence of pigment concentration can be appreciated if one relates the rate of CO_2 uptake to the chlorophyll content of the leaves (the "assimilation number" of A. Willstätter and A. Stoll)—it becomes apparent that chlorophyll is usually present in abundance. Among leaves with a green color, the effects of differences in chlorophyll content are hardly noticeable as long as irradiation is powerful, but they become evident under insufficient light—a situation found in nature in deep shade or when the sun is low. Only if there is a distinct lack of chlorophyll (as evidenced by a yellow color) does photosynthetic efficiency decline in strong light (Fig. 13). A chlorophyll deficiency sometimes occurs during development, and always in the fall when the leaves turn yellow; it also occurs in pale leaves, in which the mineral balance has been disturbed (deficiency chloroses, pp. 108–110), or in leaves subjected to desiccation, infection, or the influence of noxious gases. Finally, lack of chlorophyll can be of genetic origin, as happens in defective mutants with mottled or yellow leaves.

Pigment content is adapted to the local light climate. The shade leaves of the trees contain higher concentrations of chlorophyll than do the sun leaves. High concentrations are also found in algae grown under conditions of low light intensity. In aquatic plants an additional adaptation is possible—an adaptation to the altered spectral composition of the light at increased depth. This *chromatic adaptation* is achieved through accessory pigments, in particular the phycobilins and carotenoids. The red

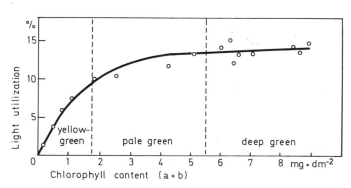

Fig. 13. Influence of the chlorophyll content of leaves upon the degree of light utilization in photosynthesis. (After Gabrielsen, 1948)

algae, which in general live at greater depths where they receive only blue-green light, absorb this light by phycoerythrin and transfer the energy to chlorophyll by way of phycocyanin.

Fixation and Reduction of Carbon Dioxide

The reactions in the presence of light capture energy and generate "reducing power" (or "assimilatory power"), the energy bound in ATP and $NADPH_2$ that is used for CO_2 assimilation. Conversion of CO_2 to carbohydrates also takes place in the chloroplasts, but in structures which lack pigment; moreover, it can proceed even in the dark. It is initiated by the binding of CO_2 to an acceptor. This compound is unstable and immediately decomposes into smaller molecules. These, finally, are converted by reduction to trioses $[(CH_2O)_3]$, from which the acceptor is regenerated. The reaction is described by the formula:

$$nCO_2 + nAcc + nNADPH_2 + nATP \xrightarrow{\text{enzyme}} (CH_2O)_n$$
$$+ nNADP^+ + nADP + nP_i + nAcc \tag{6}$$

The rate of carboxylation—that is, the speed with which the CO_2 taken up is processed—depends in particular upon the activity of the enzyme, the concentration of the acceptor, the supply of CO_2 and the temperature.

The Pentose Phosphate Pathway for CO_2 Assimilation (Calvin-Benson Cycle)

In most plants a pentose phosphate, ribulose-1,5-diphosphate (RuDP), is the CO_2 acceptor, and is decisive in determining the yield of the dark reaction of photosynthesis. Carboxylation is catalyzed by the enzyme RuDP carboxylase (= carboxydismutase). The product of this reaction, a six-carbon molecule, decomposes immediately to produce two molecules of 3-phosphoglyceric acid (PGA). Each of these molecules contains three carbon atoms, and the process is therefore also called the C_3 *pathway* of CO_2 assimilation. PGA is reduced to glyceraldehyde-3-phosphate (GAP) over several steps involving ATP and $NADPH_2$ (Fig. 14). This is the final

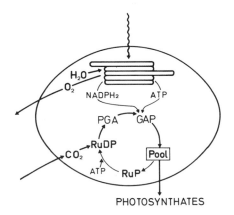

Fig. 14. Simplified diagram of CO_2 fixation and assimilation by way of the Calvin-Benson cycle in C_3 plants. RuDP, ribulose-1,5-diphosphate; PGA, 3-phosphoglyceric acid; GAP, glyceraldehyde-3-phosphate; Pool, intermediary C_3 to C_7 compounds; RuP, ribulose-5-phosphate. The photosynthates are carbohydrates, carboxylic acids and amino acids

step in raising the CO_2 taken in to the energy level of a carbohydrate. GAP flows into a pool of carbohydrates of different carbon-chain lengths (C_3–C_7), from which the acceptor is regenerated by way of ribulose-5-phosphate (RuP). The pool of carbohydrates feeds other processes of assimilation—e.g. the formation of hexoses and other photosynthetic products ("photosynthates") such as carboxylic and amino acids.

The Dicarboxylic Acid Pathway for CO_2 Incorporation (Hatch-Slack-Kortschak Pathway)

In investigations of the photosynthetic capacity of various plants, graminaceous species of tropical origin such as sugar cane, maize and millet (species of *Panicum*, *Pennisetum*, *Setaria* and *Sorghum*) had been noted for their high rates of photosynthesis. But there are also high-performance dicotyledonous plants, particularly in the genus *Amaranthus* and among the Chenopodiaceae (e.g. some species of *Atriplex* and *Kochia*, *Portulaca oleracea*, *Salsola kali*). Most of these come from hot regions that are dry for part of the year, and many are cultivated plants with strikingly high yields, or are particularly troublesome weeds. When these plants are left to assimilate in an enclosed air space under bright light, they withdraw CO_2 from the atmosphere until the concentration in the air falls below 20 ppm (the values for C_3 plants are no lower than 50—80 ppm). Highly productive plants thus can utilize a scant supply of CO_2 more effectively than the other species. They achieve this by using phosphoenolpyruvate (PEP) as the acceptor for CO_2 (Fig. 15). The binding of CO_2 to PEP is catalyzed by the enzyme PEP carboxylase, which because of its great affinity for CO_2 is effective even at very low CO_2 concentrations. The carboxylation of PEP yields oxaloacetic acid (OxAc), which is reduced to malic acid by malate

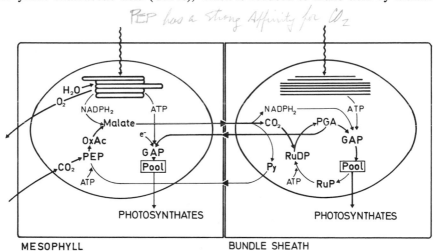

Fig. 15. A much simplified diagram of CO_2 fixation *via* the Hatch-Slack-Kortschak pathway in C_4 plants. PEP phosphoenolpyruvate; OxAc oxaloacetate; PGA 3-phosphoglyceric acid; GAP 3-phosphoglyceraldehyde; RuP ribulose-5-phosphate; Py pyruvate. PGA is also produced by carboxylation of C_2 compounds which appear in the pool; the regeneration of PEP from PGA, in which water is given off, is not shown

dehydrogenase. Oxaloacetic acid and malic acid are dicarboxylic acids, with four carbon atoms, so that this kind of CO_2 fixation has been called the "C_4 *dicarboxylic acid pathway*". The malate produced is not converted to hexose in the same cell, even though the malate-producing cells do contain the enzymes of the Calvin cycle. Rather, the intermediate product is transferred to adjacent cells specialized to utilize malate for the synthesis of larger molecules. In tropical grasses and in some dicotyledons of the C_4 type, this biochemical specialization of the tissue is even discernible in the anatomy. The malate-forming mesophyll cells have chloroplasts with normal grana-type thylacoids, they contain only a little starch, and they are arranged radially around the starch-storing bundle-sheath cells; in the chloroplasts of the latter hardly any grana structure is seen. Thus bundle-sheath and mesophyll cells are partners in CO_2 assimilation of the C_4 type. In the bundle-sheath cells, the malate received from the mesophyll cells is decomposed into CO_2 and pyruvate (Py). The hydrogen which thus becomes available is bound to a hydrogen acceptor which then becomes a reducing agent. The CO_2 released is captured by RuDP and enters the Calvin cycle. The pyruvate is sent back "empty" to the mesophyll cells, where it can serve in the regeneration of PEP. Since the bundle-sheath cells also take up CO_2 directly, PGA accumulates and can also be discharged into the mesophyll cells, where it is used to form GAP.

This apparently complicated combination of malate synthesis and the C_3 cycle gives the C_4 plants the advantage of optimal CO_2 utilization; the diversity of acceptors allows the available CO_2 and the reducing power provided by the light reactions to be used very efficiently.

CO_2 Fixation in Succulent Plants (Crassulacean Acid Metabolism, CAM)

The capacity to bind CO_2 to phosphoenolpyruvate is not unique to the C_4 plants; this reaction is widespread in the plant kingdom, but in the C_3 plants the process is insignificant compared to the pentose phosphate pathway. Many succulents, especially those of the families Liliaceae, Bromeliaceae, Orchidaceae, Cactaceae, Crassulaceae, Mesembryanthemaceae and Asclepiadaceae, display still another modification of CO_2 assimilation. They take up large quantities of carbon dioxide at night, through widely opened stomata, but of course they cannot process it photosynthetically until the next day. This phenomenon is called the De Saussure effect, or crassulacean acid metabolism (CAM). As in the C_4 plants, the CO_2 taken from the atmosphere is fixed by the conversion of PEP to OxAc and then to malate. There are in addition other reactions in which organic acids are produced by incorporation of the absorbed CO_2. These acids can be decomposed only in the presence of light, so that they accumulate during the night and are passed into the cell sap, the pH of which falls correspondingly (Figs. 16 and 19). The following morning when it is light, malate is transported back out of the vacuoles into the plastids and decarboxylated there, as in the C_4 plants. The released CO_2 is taken up by RuDP and reduced to carbohydrate. During the day, then, a normal C_3 cycle operates in the chloroplasts of succulents. The progressive emptying of the vacuoles is accompanied by an increase in the pH of the cell sap. The strikingly large vacuoles in the mesophyll of succulent leaves thus serve not only for water storage; they are also capacious

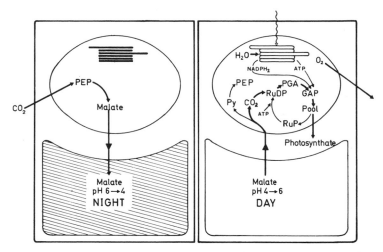

Fig. 16. Diagram of nocturnal CO_2 storage in succulents and the freeing of CO_2 from malate by day. The symbols are the same as those of Figs. 14 and 15. (Original drawing, based in part on Kluge, 1971)

storehouses of carbon that make the photosynthetic activity of the plant temporally independent of CO_2 exchange. It is ecologically advantageous to succulents—always inhabitants of dry localities—that there is a separation in time between nocturnal CO_2 fixation and further metabolism of CO_2 the following day. They are thus assured of a supply of carbon without simultaneously endangering their water balance.

Catabolic Processes

In contrast to the assimilative (anabolic) processes, through which the substance of the plant is synthesized, are those of catabolism, whereby substances are broken down to provide energy for the manifold metabolic processes of the cell. The substrates for these reactions are the carbohydrates in which energy has been incorporated by photosynthesis; in their catabolism, as CO_2 and hydrogen are released, there is a corresponding release of energy. Carbon dioxide, having lost its role as carrier of hydrogen, escapes, and the hydrogen is captured by pyridine nucleotides which thereby become charged. Most of the energy is obtained at the step in which hydrogen is transferred to the final hydrogen acceptor. In aerobic respiration this acceptor is atmospheric oxygen, in fermentation it is reducible organic compounds, and in anaerobic respiration the terminal acceptors are inorganic ions such as nitrate and sulfate. Since the water-oxygen system has a greater oxidation-reduction potential with respect to hydrogen than do other oxidizing agents, aerobic respiration provides much more energy than the other catabolic processes; it operates with an efficiency of 30—40%.

Carbohydrate Breakdown and Energy Yield in Mitochondrial Respiration

The respiratory substrate glucose is decomposed in many sequential steps: glycolysis, decarboxylation of pyruvate, the citric-acid cycle (Krebs-Martius cycle), and oxidation of the terminal electron acceptor ($NADH_2$). Glycolysis takes place in the cytoplasm, while the citric-acid cycle and respiratory-chain phosphorylation occur in the mitochondria. The entire process yields 36 moles of ATP, 2 moles of GTP, and 672-675 kcal, per mole of glucose.

Oxidative Pentose Phosphate Cycle

In addition to the breakdown of carbohydrate via glycolysis and the citric acid cycle, glucose breakdown is also possible—especially in highly differentiated plant cells—by direct oxidation, in which $NADP^+$ rather than NAD^+ is reduced. Decarboxylation in this pathway (also called the hexose monophosphate shunt) produces ribulose-5-phosphate, which is converted, in the reverse sequence, to sugars involved in the Calvin cycle. Via this *oxidative* pentose phosphate cycle and the $NADP^+$ pool, direct oxidation is closely associated with the dark reaction of photosynthesis.

Photorespiration

Another form of photosynthate breakdown—photorespiration—is even more closely related to photosynthesis than is the oxidative pentose phosphate cycle. "Photorespiration" denotes O_2 uptake and CO_2 release occurring only in photosynthetically active cells, at a rate which increases with increasing light intensity. In the dark this system is inactive. Photorespiration may well be inseparably associated with the Calvin cycle of photosynthesis, on which only the C_3 plants must depend. The C_4 plants, on the other hand, are capable of immediately refixating the CO_2 given off during respiration in the light, so that no photorespiration is apparent in these plants. Even in the C_3 plants, demonstration and especially measurement of photorespiration is difficult experimentally, because the photosynthetic gas exchange occuring simultaneously masks the respiratory gas exchange. However, in computing energy utilization (p. 81) and the photosynthetic coefficient (p. 37), one ought to know the exact level of photorespiration. Photorespiration probably occurs in cell compartments in close contact with the chloroplasts, so that it and the other forms of respiration—glycolysis, the citric-acid cycle, and the oxidative pentose phosphate cycle—exist side by side in different places in the cell.

CO_2 Exchange in Plants

The Exchange of Carbon Dioxide and Oxygen as a Diffusion Process

Carbon metabolism in the cell is linked to the external environment by gas exchange. In photosynthesis the chloroplasts use up CO_2, of which a supply must be maintained, and liberate oxygen. In parallel, by both day and night, the cells take up oxygen for respiration and give off carbon dioxide. In assimilating leaves, one or the other of these two opposed processes can predominate at a given time. The respiration occurring in the light comprises both photorespiration and mitochondrial respiration. During the day the rate of CO_2 uptake per unit of plant mass required for photosynthesis (gross photosynthesis, F_g) is greater, as a rule, than the rate at which CO_2 is freed by the total respiration in light (R_l), so that there is a net uptake of CO_2 into the leaf. Under these conditions one speaks of apparent or net photosynthesis (F_n).

$$F_n = F_g - R_l \tag{7}$$

If the rate of photosynthesis decreases it can happen that the respiration occurring simultaneously just balances it (the "compensation point" is reached). If the rate of photosynthetic activity declines still further, respiration predominates, and in the dark, respiratory release of CO_2 alone prevails. The processes participating in respiration in the dark are mitochondrial respiration and the oxidative pentose phosphate cycle.

Rates of Diffusion

Gas exchange between the cells and the surroundings of the plant (the outside air or water) occurs by diffusion. CO_2 and O_2 transport are therefore described by Fick's Law of Diffusion:

$$\frac{dm}{dt} = -D \cdot A \frac{dC}{dx} \tag{8}$$

The diffusion rate (quantity displaced, dm, in the time interval dt) depends upon the diffusion constant D and is greater the steeper the concentration gradient dC/dx in the direction of diffusion x and the greater the exchange area A. The diffusion constant depends both upon the substance considered and the medium in which diffusion takes place; in air, CO_2 and O_2 can diffuse about 10^4 times as fast as in water.

Fick's Law can be applied to gas exchange in plants in the form derived by P. Gaastra:

$$\Phi = \frac{\Delta C}{\Sigma r} \tag{9}$$

Here Φ is the flux of diffusing substance (molecules per unit area per unit time), ΔC is the concentration difference between the outside air and the site of the reaction in the cell, and Σr is the sum of a number of terms representing resistance to diffusion. Σr thus takes into account the relevant diffusion constant, effects at phase interfaces, and the spatial dimensions involved in the situation. When the equation is applied to represent CO_2 flux during illumination, for example, Φ is an expression of the net rate of photosynthesis.

The Concentration Gradient

If one assumes that when photosynthesis is proceeding briskly the CO_2 in the chloroplasts is used up as fast as it becomes available, and that during respiration in the mitochondria the O_2 concentration there falls to zero, then the concentration gradients of these two gases are determined by their concentrations in the surroundings of the plant.

The concentration of oxygen in the atmosphere is much greater than that of carbon dioxide. In terrestrial plants, therefore, respiration in the shoots is restricted by lack of oxygen only in exceptional cases, even where the diffusion resistance is high; these exceptions include massive organs not readily permeable to gases, such as thick tree trunks and large fruits. In the ground near the roots, on the other hand, oxygen deficiency is more likely. Even in the upper decimeters of the soil the oxygen content is much reduced, and at greater depths it is only half the concentration in the open air. Under unfavorable circumstances (for example, where stagnant ground water is present), the oxygen in the soil can be as little as a few volume percent. Bodies of water are always poorer sources of oxygen than the atmosphere. Depending on the temperature, not more than 14.7 mg O_2 can be in solution in a liter of water (at $0°$ C), which corresponds to a concentration by volume of 1%; at $25°$ C the solubility is only half as great.

The small concentration of *carbon dioxide*, on the other hand, means that correspondingly small concentration gradients are available. In both terrestrial and aquatic plants, under natural conditions, a suboptimal supply of CO_2 is an ever-present factor limiting the yield of photosynthesis.

Diffusion Pathway and Resistances in the Leaf

In its course from the air to the chloroplasts, carbon dioxide encounters a series of diffusion barriers. The CO_2 molecules enter the system of intercellular spaces through the stomata. In the cell walls, CO_2 moves from the gas phase, in which it diffuses relatively rapidly, to the liquid phase. The process of going into solution delays the CO_2 transport considerably. Within the cell, the dissolved CO_2 migrates slowly to the chloroplasts.

Figure 17 presents an overview of the pathways and resistances involved in transport near and within the leaf. During photosynthesis, the greatest CO_2 partial pressure is found outside a thin boundary layer of air near the leaf; the thickness of this layer depends upon the size and position of the leaves, the presence or absence of hair on the leaf surface, and above all on the degree of air movement. In still air the layer can be some millimeters thick, while a strong wind will sweep it entirely away. The thicker the boundary layer, the larger is the boundary-layer resistance r_a. If carbon dioxide is taken into the leaf more rapidly than it is replaced by diffusion, the film of air near the leaf is depleted of CO_2. CO_2 can enter the leaf only through the stomata. Some slight passage through the cuticle and the epidermal cells is demonstrable with abnormally high outside concentrations, but under natural circumstances cuticular CO_2 uptake by land plants can be discounted. The decisive constraint on CO_2 uptake into the leaf is therefore the stomatal resistance r_s; when the pores are closed r_s is nearly infinite (Fig. 18). The CO_2 concentration in the substo-

Fig. 17. CO_2 concentration gradient and transport resistances in a leaf during photosynthesis. *UE* upper epidermis; *PP* palisade parenchyma; *SP* spongy parenchyma; *LE* lower epidermis; *NPC* cells lacking chloroplasts and not photosynthetically active; *BL* boundary layer (the film of air near the leaf). During photosynthesis a *gradient in the CO_2 concentration* is established from the outside air (C_a) via the intercellular air (C_i) to the minimal concentration at the site of carboxylation (C_x). In the intercellular system of the leaf, CO_2 arrives not only from outside but also from the cells, as a result of the respiratory activity of the mitochondria (C_{RM}) and photorespiration (C_{RL}). The *transport resistances* interposed are the boundary-layer resistance r_a, the physiologically regulatable stomatal resistance r_s, diffusion resistances in the intercellular system r_i, and resistances associated with the processes of dissolving and transport of CO_2 in the liquid phase of the cell wall (r_w) and in the protoplasm (r_p). r_x indicates the "carboxylation (excitation) resistance"

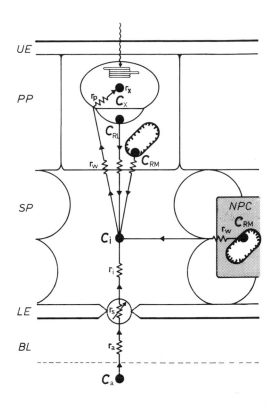

Fig. 18. Diffusion resistance for CO_2 around and within a sugar-beet leaf. The boundary-layer resistance (coarse stippling) is considerably lower than the two remaining diffusion resistances. When the pores are wide open the stomatal resistance is lower than the mesophyll resistance (fine stippling); not until the stomata narrow, as a result of insufficient light (in this case at $0.2 \cdot 10^{-2}$ W $\cdot$ cm$^{-2} \doteq 0.03$ cal $\cdot$ min$^{-1} \cdot$ cm^{-2}) does the stomatal resistance become the critical factor limiting diffusion. (After Gaastra, 1959)

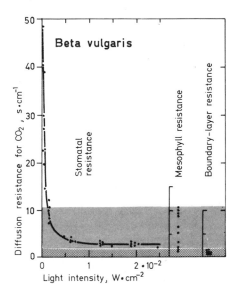

matal cavity and in the intercellular air C_i is already much lower than that in the outside air (C_a), but it is still appreciable. The intercellular air not only serves as a source of CO_2, it also receives CO_2 as a result of respiratory processes in green and non-green cells. A *"compensation point" in the gas-exchange equilibrium* can thus be defined; this reflects the state in which C_i is equal to C_a. In this situation, there is no net gas exchange even though the stomata may be wide open. The diffusion resistance in the intercellular system (r_i) depends upon the structure of the leaf. As a rule the mobility of gas in shade leaves is greater than in sun leaves, and greater in soft leaves than in more sturdy assimilation organs. Two other resistive factors of about the same order of magnitude as the intercellular diffusion resistance are the interface resistance r_w, associated with the transition from the gas to the liquid phase at the cell walls, and the diffusion resistance r_p within the protoplasm and the chloroplasts.

The path of CO_2 transport ends in the chloroplasts with fixation by the acceptor. The speed with which CO_2 is processed also affects the steepness of the concentration gradient and thus the influx of CO_2. The rate of carboxylation, in comparison with the light reactions of photosynthesis, can be sluggish and become a rate-limiting "bottleneck" in the overall process. This fact is sometimes expressed in the literature with reference to a "carboxylation (or excitation) resistance" r_x, though this resistance is not actually associated with the movement of CO_2. Because PEP carboxylase catalyzes CO_2-binding much more effectively than does RuDP carboxylase, the "carboxylation resistance" in C_4 plants is much smaller than in C_3 plants.

Regulation of Gas Exchange by the Stomata

Opening of the Stomata

The apparatus controlling the size of the slits through which gas enters the leaves is the most important regulator in the diffusion process. By varying the width of the opening, the plant simultaneously regulates both CO_2 entry into the leaf and water loss by transpiration.

Number, distribution, size, shape and mobility of the stomata are species-specific characteristics, though they vary according to habitat and even among individuals (Table 2). The critical anatomical dimension determining stomatal resistance is the *pore width*. The maximal width to which the pore of a stoma can be opened, which depends upon the shape and the properties of the walls of the guard cells, limits the rate at which gas can flow through; maximal width corresponds to minimal r_s. The opening capacity is greatest in the leaves of herbaceous dicotyledons and the foliage of trees with open crowns, somewhat less in the leaves of the other soft-foliage trees and the grasses, and still less in the evergreen woody plants with thick, stiff leaves.

The extent of gas exchange via the stomatal pores depends on the *total pore area* per unit leaf area. This ratio is computed as the product of pore density (number of stomata per mm² leaf surface) and maximal area per pore. The pore-area ratio thus indicates the maximal cross section through which gas exchange can occur. In most plants the pore area comprises 0.5–1.5% of the leaf surface, though in exceptional cases it can be as much as 3%. The evergreen sclerophyllous plants of the maquis

Table 2. Stomatal density, pore width and pore area in various plants. (From Stocker, 1929; Meidner and Mansfield, 1968; Pisek *et al.*, 1970; Cintron, 1970)

Plants	Stoma density (number per mm^2 leaf surface)	Pore length (μm)	Maximal pore width (μm)	Relative pore area (% of leaf surface)
Grasses	(30) 50—100	20—30	ca. 3	0.5—0.7
Herbaceous sciophytes	40—100 (150)	15—20	5— 6	0.8—1.2
Herbaceous heliophytes	100—200 (300)	10—20	4— 5	0.8—1
Winter-deciduous foliage trees	100—500	7—15	1— 6	0.5—1.2
Tropical forest trees	20— 60	10—20	5—10	0.2—0.6
Evergreen sclerophylls	100—300	10—15	1— 2	0.2—0.5
Conifers	40—120	15—20	—	0.3—1

and dwarf-shrub heath have an especially small pore area, corresponding to the limited opening capacity of the stomata.

Factors Influencing Stomatal Behavior

The degree of opening of the pores, and thus the stomatal diffusion resistance, depends upon both the environment and the interior state of the plant. The most influential external factors are light, temperature, humidity and water supply; the internal factors include the CO_2 partial pressure in the intercellular system, the content of water and ions in the tissues, and phytohormones; the hormones gibberellic acid and cytokinin promote opening, while abscisic acid promotes closing.

In the light, with an adequate supply of water, the stomata open wider the greater the **light intensity.** M.G. Stålfelt, who investigated thoroughly the factors influencing the mechanics of the stomata, called this opening mechanism "photoactive" opening. A second factor crucial to the degree of opening of the pores is **carbon dioxide concentration.** Its influence is evident in the dark. In the dark, at a CO_2 concentration in the air of 300 ppm, the pores are closed, but they open if the CO_2 concentration is reduced. It makes no difference which side of the epidermis is subjected to the low concentration of CO_2. In the light, the consumption of CO_2 by photosynthesis causes the CO_2 pressure in the intercellular spaces to fall, so that "photoactive opening" is brought about not only by a direct action of light but also indirectly by way of CO_2 depletion. One could just as well speak of a "chemoactive" component in pore regulation. The decisive influence of CO_2 concentration upon pore width is especially apparent in the succulents with crassulacean acid metabolism; they open the pores during the night when the partial pressure of CO_2 in the intercellular system falls as a result of the intensive formation of malate, and they close the pores when the breakdown of malate frees CO_2, which accumulates in the intercellular system until it is further metabolized (Fig. 19). Chemoactive responses of the stoma-

31

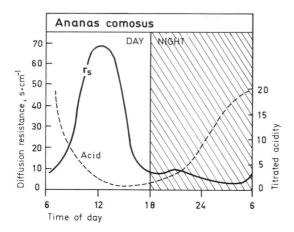

Fig. 19. Daily rhythm of the stomatal diffusion resistance and the acidity of the cell sap of pineapple leaves. In the morning malate is decomposed (the acidity falls), freeing CO_2. This raises the CO_2 content of the intercellular air, and the stomata respond to this chemical stimulus by closing (the steep rise of r_s). In the course of the day CO_2 is withdrawn from the intercellular air by photosynthesis and the stomata open again. During the night CO_2 is used up in the formation of malate (the acidity rises), the CO_2 content of the intercellular air remains low, and the pores are therefore open (r_s low). Acidity was determined by titration and is given in terms of the number of ml of 0.1N NaOH required to neutralize 5 ml of cell sap. (After Aubert, 1971)

ta (usually closures) are caused not only by CO_2 but also by other substances—e.g., SO_2 in the atmosphere and the fungicide phenylmercury acetate, which thus finds use as an antitranspirant. Toxins released by parasitic fungi (*Helminthosporium maydis*) also cause stomatal closure.

The **temperature** influences primarily the *speed* of opening, which is dependent upon the energy available for carrying out this movement. At higher temperatures more such energy can be provided, so it is not surprising that the opening mechanism operates more rapidly as the temperature is increased. At lower temperatures (below about 5° C) the pores open very slowly and incompletely, and when the temperature falls to levels of 0° to −5° C they remain closed (Fig. 20). In extreme heat, however, the stomata also close.

Another important environmental factor influencing the pore width is **water**. The pore width increases with the turgidity of the guard cells, and as turgidity decreases, the gap closes. Stålfelt called this process "hydroactive closing". This was not the best choice of words, however, since the loss of turgor is not an active process, but is brought about by passive water loss. Closing of the pore is thus the consequence of a poor water supply to the leaf. There are indications that the guard cells, in some species, at least, transpire more strongly than the other epidermal cells. In such cases a lack of water is reflected sooner in the region of the stoma than elsewhere in the leaf; the closing reaction, which takes 10—15 min, therefore occurs in good time when there is danger of desiccation.

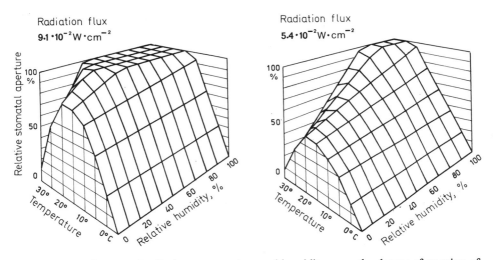

Fig. 20. The influence of radiation, temperature and humidity upon the degree of opening of the stomata in the leaves of *Ligustrum japonicum*. The degree of opening was determined by porometer measurement. (After Wilson, 1948)

Given the combined effects of all the foregoing external factors, the pore width is usually set not at the maximal but at an intermediate value, since it is rare that all the conditions promoting opening are present at once (Fig. 20). Similarly, complete closing is induced only under markedly unfavorable circumstances, which also occur only rarely.

Regulation of Stomatal Movement

The extremely sensitive regulation of stomatal diffusion resistance can be thought of as comprising two control systems, the CO_2 control system and the H_2O control system. The CO_2 system responds to the partial pressure of carbon dioxide in the intercellular region and acts to maintain a favorable CO_2 concentration in the leaf; *demand regulates supply*. The CO_2 control system reacts within minutes, i.e. with extraordinary promptness; but it dominates only if the water content of the leaf is adequate. When the water supply is restricted, the H_2O control system takes over. It reacts to changes in humidity as well, and determines pore size via a direct effect upon the turgidity of the guard cells.

The physiological mechanism of guard-cell movement is not yet completely explained, and there are a number of hypotheses to be considered. A reliable starting point is the fact that the movements are brought about by changes in turgor. When the cells are flaccid, the pore is closed, and the process of opening requires energy. The increase in turgor can be elicited by active ion transport (chiefly of K^+ and phosphate), by changes in permeability, and by enzymatic processes. Further factors involved are the availability of ATP, the processes of carboxylation, and the enzymatic conversion of starches into sugar and back again.

Photosynthetic Capacity and Specific Respiratory Activity

Photosynthetic Capacity

Under optimal conditions, i.e. strong illumination, good water supply, favorable temperature, and air in which the CO_2 concentration is raised artificially to 0.3–1% by volume, herbaceous plants are capable of taking up as much as 150 mg CO_2 per dm^2 leaf surface per hour. At this point the limit of the carboxylation process is reached. Such maximum performance ("potential photosynthesis") is possible only under laboratory conditions and probably only for short periods, since otherwise excessive amounts of photosynthates would accumulate in the cells. In nature, where the CO_2 content of the atmosphere rarely exceeds 0.03–0.04% by volume, the plants are capable of binding at best 80 mg $CO_2 \cdot dm^{-2} \cdot h^{-1}$ (cf. also Fig. 34).

The maximum rate of net photosynthesis by a plant at a given state of development and activity, under natural conditions of atmospheric CO_2 content and optimal conditions with respect to all other external factors, in the terminology of A. Pisek and W. Larcher, is called its *photosynthetic capacity*.

Photosynthetic capacity is a quantity measured under standard conditions; it can be used to characterize certain physiological types of plants as well as plant species, ecotypes and even individual varieties. Within the plant kingdom there are pronounced differences in photosynthetic capacity: A survey is given in Table 3. As is to be expected, the C_4 plants take the lead. Next come agricultural plants that accomplish CO_2 fixation via the Calvin cycle; the high rates of photosynthesis in these plants are due in great degree to successful breeding. Among these crops rice, wheat, potatoes and the sunflower show particularly high performance. The thallophytes are at the bottom of the list. The C_4 plants take up 30 times as much CO_2 as do mosses, lichens and algae, and they assimilate CO_2 with almost double the yield of most of the agricultural plants. Among the vascular plants, herbs outperform woody plants, and in both the herbaceous and woody plants those forms adapted to shade manage only half to a third of the carbon utilization of plants growing in sunny locations. Species with assimilation organs of small surface area, like grasses with rolled leaves, dwarf ericaceous shrubs with grooved leaves, the needles of conifers, shrubs with assimilating shoots, and succulents, capture little of the incident light and thus display only moderate assimilative activity. The aquatic plants form a distinct group with a surprisingly small capacity to bind CO_2. Even if one disregards the values for planktonic algae, which are difficult to apply comparatively, a clear deficiency in photosynthetic capacity remains—even with respect to the submersed vascular plants. A primary reason for this is thought to be that sessile aquatic plants are not as well supplied with CO_2 as land plants. It is true that fresh water contains about 160 times as much CO_2 as the air; but the rate of diffusion of CO_2 in water is only about 10^{-4} that in air, so that the CO_2 supply underwater at the surface of the leaves is restricted.

The differences in photosynthetic capacity among species and varieties are of ecologic significance, but their greatest importance is as a basis for the selective breeding and cultivation of plants valuable in agriculture, gardening and forestry. The causes of these specific differences are found in the effectiveness of the enzymes as

Table 3. Average maximum values for net photosynthesis under conditions of natural CO_2 availability (300 ppm), saturating light intensity, optimal temperature and an adequate water supply. (From measurements by numerous authors)

Plant group	CO_2 uptake	
	mg $\cdot$ dm^{-2} $\cdot$ h^{-1}[a]	mg $\cdot$ g$^{-1}W_d$ $\cdot$ h^{-1}[b]
A. Land plants		
1. Herbaceous flowering plants		
Plants with CO_2 fixation by C_4 pathway	50—80	60—140
Agricultural C_3 plants	20—40	30—60
Herbaceous heliophytes	20—50	30—80
Herbaceous sciophytes and spring geophytes	4—20	10—30
Grasses	6—12	
Desert plants	4—12	2—8
2. Succulents (CAM plants)		
Light	3—20	0.3—2
Dark CO_2 uptake	10—15	1—1.5
3. Woody plants		
Winter-deciduous trees and shrubs		
sun leaves	10—20 (25)	15—25 (30)
shade leaves	5—10	
Evergreen broad-leaved species of the tropics and subtropics		
sun leaves	8—20	10—25
shade leaves	3—6	
Sclerophyll species from regions with summer dry season (maquis, bushland)	5—15	3—10
Evergreen conifers	4—15	3—18
Dwarf shrubs with grooved leaves	4—10	4—6
4. Cryptogams		
Ferns	3—5	
Mosses	*ca.* 3	2—4
Lichens	0.5—2	0.3—2 (3)
B. Aquatic plants		
Swamp plants	20—40	
Submersed tracheophytes	4—6	*ca.* 7
Planktonic algae		*ca.* 3

[a] To allow comparison of photosynthetic capacity of different plant types, the photosynthetic rates are normalized per unit surface area. The surface area is that area receiving radiation, not the total area of upper and lower surfaces.
[b] The photosynthetic rate per unit dry weight of leaf; this number can be used to calculate the length of time required for a leaf to acquire the carbon necessary to form another leaf of a given mass.

well as in anatomical peculiarities of the leaf structure, the ease with which air passes through the intercellular system, and the shape and distribution of the stomata.

Respiratory Activity

The rate of respiratory activity in a plant species differs from one organ to another, and it changes both with the state of development and activity and with the temperature. To facilitate comparison, the *specific respiratory activity* is expressed as the rate of respiration measured in the dark at a standard temperature, usually 20° or 25°C (Table 4). Herbaceous species, especially those with a rapid growth rate, respire twice as rapidly as the foliage of deciduous trees under the same conditions, and the latter in turn respire on the average at five times the rate of the assimilation organs of evergreens. Within a given group, heliophytes respire distinctly more rapidly than sciophytes. Different species of plants exhibit characteristic differences in respiratory activity; ratios may be of the order of 1:10 to 1:20. In a single plant, flowers and unripe fruits respire at a greater rate than leaves, and roots more rapidly than the axial parts of the shoots. In branches and tree trunks the primary respiratory regions are the bark, the cambium, and the outermost cell layers of the woody tissue. When expressed with respect to the periderm area (gas exchange per unit surface area), the respiratory activity measured at a given temperature in twigs, branches and tree trunks of different thicknesses increases with diameter (Fig. 21); when expressed with respect to weight (exchange per unit mass) it decreases, because the woody axes of shoots contain much non-respiring material, the proportion of which steadily increases with increasing diameter.

Table 4. Respiration of mature leaves in the dark in summer at 20° C. (From the measurements of numerous authors)

Plant group	CO_2 release $mg \cdot (gW_d \cdot h)^{-1}$
Crop plants	3—8
Wild herbs	
Heliophytes	5—8
Sciophytes	2—5
Winter-deciduous foliage trees	
sun leaves	3—4
shade leaves	1—2
Evergreen foliage trees	
sun leaves	ca. 0.7
shade leaves	ca. 0.3
Evergreen conifers	
sun needles	ca. 1
shade needles	ca. 0.2

Fig. 21. Respiration of larch branches and trunks of different thicknesses. As the diameter increases the proportion of non-respiring woody material rises, so that the value for respiration per unit dry weight (R_{weight}) decreases with increasing diameter; the CO_2 given off per unit surface area of shoot (R_{area}) increases only slightly from an intermediate thickness on. (After Tranquillini and Schütz, 1970)

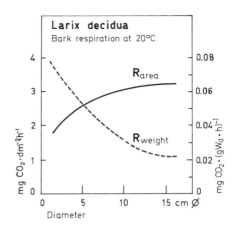

The Photosynthetic Coefficient

The net yield of photosynthesis is greatest when high photosynthetic capacity is paired with moderate respiration. This situation can be quantified in terms of the coefficient

$$k_F = \frac{F_g}{R} \doteq \frac{F_n + R_d}{R} \tag{10}$$

where, as before, F_g = gross photosynthesis, F_n = net photosynthesis, R_d is dark respiration, and R is total respiration. The photosynthetic coefficient k_F indicates how much the leaf organs must divert from the overall photosynthetic intake for their own respiration, under the most favorable conditions. Since F_g is not directly measurable, it is customary to approximate it (as in Eq. 10) by the sum of net photosynthesis and dark respiration (at a corresponding temperature). k_F is useful in comparing the yield obtained in the gas exchange process in different plants. The leaves of vascular plants can bind 10–20 times as much CO_2 as they use for respiration in a given period. The largest values of k_F are measured for the sun leaves of trees rather than for the cultivated herbs. In such leaves the intake of CO_2 is rapid, and there is thrifty consumption of carbohydrates as well. In the latter respect, they outperform herbaceous plants. Photosynthetic coefficients are particularly low in conifer needles (between 4 and 8) and in lichens (less than 5), which is understandable in view of the fact that both contain a large proportion of tissue engaged only in respiration, and not in photosynthesis.

Influence of Developmental Stage and Activity State upon Respiration and Photosynthetic Capacity

Photosynthetic capacity and respiratory activity are characteristic of a plant species, but they are not constants. Within a given plant, gas-exchange behavior changes in the course of development and with seasonal and even diurnal fluctuations in activity.

Younger plants *respire* more rapidly than older plants. In the growing parts of plants, respiration is particularly prominent; for the extensive synthesis of new tissue, ATP is hydrolyzed at a rate far exceeding normal operating requirements. A feedback mechanism enables the cell to speed up respiratory ATP formation according to demand. In seedlings, at the tips of roots, during leafing-out and in developing fruits, the respiration that supports synthesis amounts to between three and ten times the normal operating respiration. With increasing differentiation and maturation of the tissues, respiratory activity returns to a much lower level (Fig. 22). The onset of the breakdown processes of aging in the leaves, and especially in the fruits of some plant species, may be presaged by a transient sharp increase ("climacteric rise") in respiration. This is a sign of disturbed metabolism, also recognizable in the discoloration of the foliage and the release of gaseous metabolic products (for example, ethylene) by fruits.

Photosynthetic capacity also changes in the course of development. Very young leaves are not yet able to fix CO_2 efficiently; particularly in conifers, photosynthetic activity is initially so slight that it is exceeded by the very intensive, simultaneous respiration associated with synthesis of new tissue. When the new shoots of woody plants are being extended, therefore, one can measure a net CO_2 release all day long in the light; the young shoots cannot maintain themselves and must be supplied with carbohydrates from older parts of the plant (Figs. 23 and 25). As the leaves unfold, however, the capacity for intensive CO_2 fixation develops rapidly. Young, fully developed foliage is at a peak in this respect, and only days or weeks later photosynthetic capacity begins to decline, falling steadily as the plant grows older. This decline occurs more rapidly in assimilation organs that are functional for only one season, whereas in foliage that lasts for several years it is a slower, interrupted process. In evergreen plants from regions with a cold season, photosynthetic capacity falls off after every winter and each time new leaves are formed (Fig. 23). If the formation of new leaves is omitted, or if one removes the new growth, the life span of the old leaves is extended and the process of aging, with respect to photosynthetic capacity, is slowed.

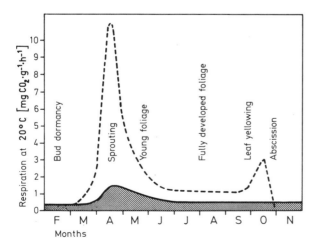

Fig. 22. Respiration of leaves of deciduous trees (dashed line) and evergreen woody plants (stippled area) as a function of the state of development of the vegetation. (Based on the data of Eberhardt, 1955; Pisek and Winkler, 1958; Neuwirth, 1959; Larcher, 1961; Negisi, 1966; E. D. Schulze, 1970)

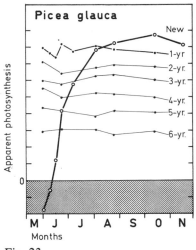

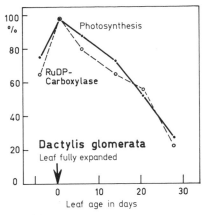

Fig. 24

Fig. 23

Fig. 23. Dependence of photosynthetic capacity of a spruce upon the age of the needles. The newly-sprouted needles develop their full photosynthetic capacity only gradually; during the expansion phase the rate of photosynthesis is low and respiration is much increased, and as a result they give off CO_2 even in the light (stippled area = CO_2 release). In subsequent years the photosynthetic capacity of the needles sinks a little lower after each winter that the needle survives. (After Clark, 1961)

Fig. 24. Decline of photosynthetic capacity and activity of RuDP carboxylase with increasing age of the leaves in orchard grass. (After Treharne and Eagles, 1970)

The dependence of photosynthetic capacity on the stage of development has several causes. Very young leaves have not yet acquired their full surface area and capture correspondingly less light; furthermore, they are usually low in chlorophyll and respire at a high rate. The chief effect, however, is due to changes in the activity level of the enzymes during development (Fig. 24), which in turn are controlled by phytohormones. Differences in enzyme activity may also account for the fact that in some species, e.g. in *Calluna*, *Citrus*, and various herbs, the maximum photosynthetic performance is shifted to the generative period of life (the flowering phase, fruit formation). In the final stages of the life cycle, shortly before the plant dies back or the leaves fall, photosynthetic capacity collapses entirely for the additional reason that chlorophyll is broken down or, in grasses for example, the stomata stiffen.
Photosynthetic capacity and respiratory activity also change with the alternation between the active and dormant periods of plants. During the period of *winter dormancy* the assimilation rate can fall to zero for weeks, with a simultaneous decrease in respiratory activity—an indication of the relative suspension of overall metabolism of the plant (Fig. 25). In regions with a mild winter the reduction of photosynthesis is not so dramatic, but it is distinctly depressed (cf. Fig. 38).

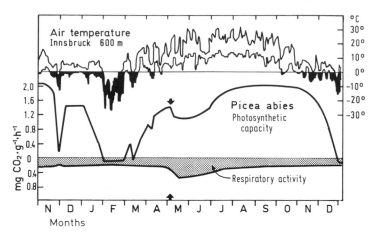

Fig. 25. Photosynthetic capacity and respiratory activity of the needle-bearing tips of twigs of a mature spruce. The rate of respiration is shown as the stippled band, increasing downward. The twigs were gathered outdoors and investigated in the laboratory under standard conditions (photosynthesis at 12° C and 10 kLx, respiration in the dark at 20° C). The effect of outdoor air temperature persists in the laboratory; photosynthetic capacity is reduced as soon as the daily minima of the air temperature fall regularly below 0° C, and if the temperature maxima also remain below freezing, net photosynthesis is completely suspended. When the new shoots appear (arrows) there is a transitory decline in net photosynthetic capacity as a result of the increased respiration of the growing tips of the twigs. (After Pisek and Winkler, 1958)

The Effect of External Factors on CO_2 Exchange

CO_2 exchange is influenced by a number of external factors. As a photochemical process, photosynthesis is of course directly dependent upon the availability of radiation. The dark reactions of photosynthesis and respiration are purely biochemical processes, influenced especially by temperature and the supply of CO_2. A given environmental factor can affect the system at several points. For example, radiation serves primarily as an energy source for the photochemical reaction, but it also exerts indirect effects upon the degree of photosynthetic activity and CO_2 exchange in other ways. These include its setting of the stomatal opening, its effect on the temperature of the plant and its role in photorespiration.

The Dependence of Net Photosynthesis on Light

The Light-Dependence Curve

If leaves or suspensions of algae are exposed to increasing intensities of illumination, the CO_2 uptake increases at first in proportion to light intensity and then more slowly to a maximum value. That is, the relationship between net photosynthesis and radiation is represented by a *saturating curve*. This light-dependence curve in dim light reflects a net release of CO_2, since more CO_2 is given off by respiration than is

Fig. 26. Dependence of CO_2 exchange upon light intensity, in sun and shade leaves of the beech. Measurements were made at 30° C. The region of weak light (the section enclosed by the box in the upper drawing) is shown in the lower drawing with the abscissa expanded. Leaves adapted to shade respire at a lower rate than those adapted to light; they reach the light compensation point (I_K) at a lower intensity. Moreover, in the region of the curve between I_K and I_S, they utilize the light more efficiently, but I_S is at a lower intensity. (After Retter, 1965)

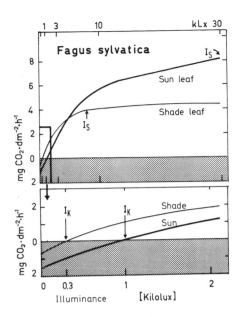

fixed by photosynthesis (Fig. 26). At somewhat greater intensities the light compensation point is reached. At the *compensation light intensity* I_K photosynthesis fixes exactly as much CO_2 as is set free by respiration. Plants that respire rapidly thus require more light for compensation than do those with slower respiration rates. Once the compensation point has been passed CO_2 uptake increases rapidly. In the lower range of this increase there is a strict *proportionality* between the yield of photosynthesis and the available radiation. The speed of the light reactions is the limiting factor for the overall process in this range. With very high light intensity the yield of photosynthesis continues to increase only slightly or not at all; the reaction is *light-saturated* at this point (I_S) and the rate of CO_2 uptake is now limited not by photochemical but rather by enzymatic processes, and by the supply of CO_2.

A comparison of the light-dependence curves of different plant species (Fig. 27) shows that the species that fixate CO_2 via the dicarboxylic acid pathway stand out. C_4 plants such as millet and maize are not light-saturated even at the highest intensities, and even at intermediate irradiance they operate more efficiently than the C_3 plants. Evidently PEP carboxylase, even at the strongest light intensities applied here, is capable of keeping pace with the light reaction. The Calvin cycle of the C_3 plants is much less efficient, and thus the light-dependence curves saturate at lower intensities. There are even plants in which photosynthetic performance falls off under excessive illumination, so that the curve shows an intensity optimum. Most of these are cryptogams, but tree seedlings and herbs growing in the underbrush of dense woods also exhibit this behavior.

The positions of the cardinal points I_K and I_S reflect the light conditions in the natural habitat of the plants and characterize the different kinds of plants (Table 5 and Fig. 26). Leaves that have grown in the shade respire less than sun leaves and therefore compensate at considerably lower light intensities. In general, the I_K of

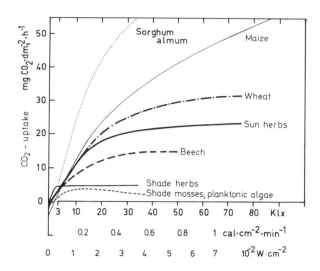

Fig. 27. Light-dependence of net photosynthesis in various plants at optimal temperature and with the natural supply of CO_2. (After Stålfelt, 1937; Böhning and Burnside, 1956; Gessner, 1959; Retter, 1965; Stoy, 1965; Hesketh and Baker, 1967; Ludlow and Wilson, 1971 a, b, c; and the results of numerous other authors)

Table 5. Light-dependence of net photosynthesis of single leaves, under conditions of natural CO_2 availability and optimal temperature. (From the measurements of numerous authors)

Plant group	Compensation light intensity I_K, in kLx	Light saturation I_S, in kLx
A. Land plants		
1. Herbaceous plants		
C_4 plants	1—3	over 80
Agricultural C_3 plants	1—2	30—80
Herbaceous heliophytes	1—2	50—80
Herbaceous sciophytes	0.2—0.5	5—10
2. Woody plants		
Winter-deciduous foliage trees and shrubs		
Sun leaves	1—1.5	25—50
Shade leaves	0.3—0.6	10—15
Evergreen foliage trees and conifers		
Sun leaves	0.5—1.5	20—50
Shade leaves	0.1—0.3	5—10
3. Mosses and lichens	0.4—2	10—20
B. Water plants		
Planktonic algae		(7) 15—20

shade plants lies at 0.5—1% of full sunlight. Shade leaves furthermore utilize weak light better than do the sun leaves and reach their light-saturation point at very low intensities, near 10 kLx (about 0.1—0.3 cal · cm^{-2} · min^{-1}). Heliophytes, inferior to sciophytes in twilight, make better use of bright light and thus produce a significantly higher photosynthetic yield. Agriculturally important plants, which should produce as large yields as possible, must therefore be heliophytes.

The variability in light dependence of photosynthesis among different types of plants can also be seen under the conditions prevailing in the natural habitat. If gas exchange is not restricted by other local environmental factors such as the water

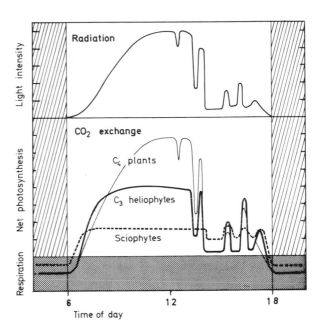

Fig. 28. Schematic diagram of the daily fluctuation in CO_2 exchange as a function of the available radiation. *C_4 plants* can utilize even the most intense illumination for photosynthesis, and their CO_2 uptake follows closely the changes in radiation intensity. In *C_3 plants* photosynthesis becomes light-saturated sooner, so that strong irradiation is not completely utilized. *Sciophytes*, adapted to utilization of dim light, take up more CO_2 in the early morning and late evening, as well as during periods when the sun is obscured, than do the heliophytes; but the former do not utilize bright light as efficiently

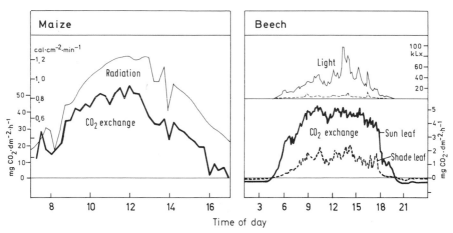

Fig. 29. Variations in net photosynthesis measured throughout the day as a function of the available radiation. In maize (a C_4 plant), photosynthesis follows the daily fluctuation in radiation. In sun leaves of the beech, F_n follows the changing illumination only up to about 50 kLx, and in shade leaves, F_n undergoes short-term fluctuations associated with the brief fluctuations in brightness. (After Hesketh and Baker, 1967; E. D. Schulze, 1970)

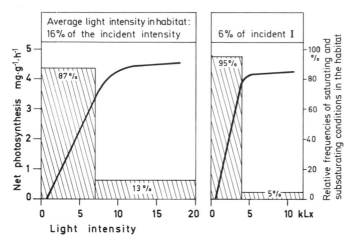

Fig. 30. Light-dependence of net photosynthesis of oak saplings in the undergrowth of a wood. *Left*: plants that have developed at an average relative intensity of 16%. *Right*: plants adapted to deeper shade, having grown at an average relative intensity of 6%. The graphs indicate that under the tree canopy the undergrowth is provided with saturating light during only 5% or 13%, respectively, of all daylight hours. During the remaining hours of the day the light intensity is in the range of proportionality of the light-dependence curve; in this region the available radiation is fully utilized. (After Malkina *et al.*, 1970)

supply and the temperature, net photosynthesis parallels light availability up to the saturation region. In the case of C_4 plants, this means that they can make full use of the light at noon on a clear day (Figs. 28 and 29). In C_3 plants the increase of photosynthetic activity ceases at the time of day when the irradiation exceeds I_S. Bright passing clouds have little effect on the rate of photosynthesis of heliophytes, but there is an effect of the more marked fluctuations in illumination caused by variable cloud cover. The sciophytes on the woodland floor and the shade leaves in the interior of tree crowns are affected by the variations in the sunlight penetrating the foliage only if illumination stays below about 10 kLx. But little is lost by their inability to follow at higher intensities, for beneath a moderately dense tree canopy (average available radiation in the herbaceous layer is 6—16%) the light intensity is above their saturation point for only about $^1/_{10}$ of the time during the daylight hours (Fig. 30).

Light-Dependence of Photosynthesis within a Stand of Plants

Observations of single leaves could lead to the mistaken inference that there tends to be a surplus of light. But for the plant as a whole and for stands of plants this is not so. It is true that the individual leaves of a plant are often arranged so as to favor interception of the strongest average light, but leaf orientation is seldom perpendicular to the direct incident radiation from the sun. In the course of the day the leaves of a plant are struck by light at many different angles, and only rarely are they exposed to the full incident radiation.

Fig. 31. Net photosynthesis in perpendicularly illuminated oat leaves and in oat plants with the natural upright orientation of the leaves, as a function of light intensity. Because the light strikes them at an angle and they partially shade one another, the leaves in a natural stand of plants are not saturated with light even when the incident light is relatively intense. (After Boysen-Jensen, 1932)

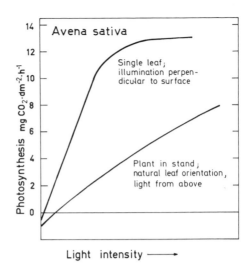

In a stand of plants, the contribution of the various layers of foliage to the overall photosynthetic yield is quite different, depending on the arrangement and the amount of shading of one layer by another. In the morning, in a field of alfalfa 30 cm high, the light compensation point is exceeded two hours later in the lowest layer of leaves than in the top layer of leaves, and even under strong sunlight the bottom layer displays only 3% of the photosynthetic activity of fully illuminated leaves. Even the layer of leaves at half height (10—20 cm above the ground) attains little more than 10% of full performance. The situation is similar within the dense crowns of trees and beneath the tree canopy of a forest. There, too, light saturation of photosynthesis is possible only in the outermost and uppermost regions, while within the crown net photosynthesis at noon falls to 15% or less of the possible value. The compensation point in such shaded regions is reached only in late morning, and the illumination falls below it again several hours before sunset. The photosynthetic performance of a stand of plants as a whole rises gradually with increasing irradiation, and continues to increase even after the light-saturation points of the outer leaves have been exceeded (Fig. 31). Because of the extensive overlapping of the foliage, whereby the shaded leaves are still unsaturated at the intensities incident upon them, even the highest intensities occurring in nature are utilized.

Photosynthesis by Planktonic Algae

In bodies of water, within the euphotic layer colonized by phytoplankton, there is a characteristic gradual variation of photosynthetic activity with depth—even though the illumination gradient is approximately exponential (Fig. 32). The photosynthetic depth profile is associated with the shape of the light-dependence curve for algae, which instead of saturating passes through an optimum and then falls (cf. Fig. 27). The highest rates of photosynthesis when the sun is high are thus measured not directly below the surface of the water but somewhat deeper, between 2 and 15 m (2—5 m in lakes, 10—15 m in the open ocean); the effect of course depends on the

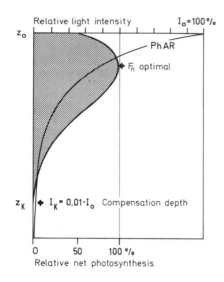

Relative light intensity $I_0 = 100\,\%$

PhAR

F_n optimal

z_0

z_K $\leftarrow I_K = 0.01 \cdot I_0$ Compensation depth

0   50   100 %
Relative net photosynthesis

Fig. 32. Photosynthetic activity of phyto-plankton and relative light intensity, as a function of depth under water. At the water surface (z_0) photosynthesis is supersaturated with respect to light when the sun is high, and therefore is somewhat reduced; a few centimeters to decimeters beneath the surface the illumination becomes optimal; and at the compensation depth z_K, where the light intensity is only 1% of that at z_0, net photosynthesis falls to zero. (After Talling, 1970)

intensity of the light and the turbidity of the water. On overcast days and during the seasons of diminished radiation the light does not reach above-optimal intensities, and the region of maximal photosynthesis is shifted up to the water surface. At greater depths, photosynthetic activity declines until eventually it is just able to compensate for respiration. Compensation as a rule is found at a depth (the compensation depth) at which not more than 1% of the surface radiation penetrates.

Temperature-Dependence of Photosynthesis and Respiration

The effect of temperature upon the processes of photosynthesis and respiration is exerted through the temperature dependence of the various enzymes involved. A general description of chemical reactions is given by the Van't Hoff equation relating reaction speed k and absolute temperature T. Since in this equation k is proportional to an exponential function of $1/_T$, it is approximately the case, over the small temperature range of biological reactions, that a given increment in T (say, a $10°$ increase) will increase k by a constant multiple. This multiple, for a $10°$ increase, is denoted Q_{10}; for many reactions it has a value of about 2. That is

$$Q_{10} = \frac{k_{T+10}}{k_T} \doteq 2 \tag{11}$$

Temperature-Dependence of Carbon Binding

Temperature influences photosynthesis only through the dark reactions; the photochemical process is nearly independent of temperature. The fixation and reduction of carbon dioxide occurs with increasing speed as the temperature rises, until a maximum value is reached; this rate is then maintained over a broad range of temperatures. Only at considerably higher temperatures, when the enzymes begin to be inactivated and the interplay of the various reactions is disturbed, does photosynthe-

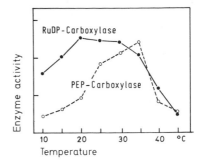

Fig. 33. Temperature dependence of the activity of RuDP carboxylase from grasses of the temperate zone (C_3 plants) and of PEP carboxylase from tropical grasses (C_4). (After Treharne and Cooper, 1969)

sis come sharply to a halt. The activities of the enzymes crucial to CO_2 binding, RuDP carboxylase and PEP carboxylase, show a distinct—and species-specific—dependence upon temperature (Fig. 33). PEP carboxylase appears to be less efficient at low temperatures than RuDP carboxylase.

Temperature-Dependence of Respiration

As the temperature rises, in the region of 5—20° C, respiration in the dark increases with a Q_{10} of 2. Below 5° C the Q_{10} is greater, and above 20° C it falls slowly to 1.5 or less. Considerable changes in respiration can be observed when cold conditions are imposed. The respiratory metabolism of cold-sensitive tropical plants becomes severely disorganized even at temperatures between 0° and 5° C.

Limiting Temperatures and the Temperature Optimum for Net Photosynthesis

The temperature-dependence of net gas exchange results from the difference between the rate of photosynthetic CO_2 incorporation and the rate of respiration prevailing at a given temperature (Fig. 34). There is an additional effect of temperature upon the degree of stomatal opening (cf. p. 32). Net photosynthesis is measured over a range in which increasing temperature has a stimulative effect, and over another in which the effect is inhibitory. These regions are defined by three *cardinal points*: the cold limit or the temperature minimum (T_{min}) for net photosynthesis, the temperature optimum (T_{opt}), and the heat limit or the temperature maximum (T_{max}) for net photosynthesis.

Optimal Range of Temperature. That range of temperature in which net photosynthesis is more than 90% of the maximum obtainable can be regarded as optimal. The temperature optimum for *net* photosynthesis is narrower than the optimal temperature span for the activity of the photosynthetically important enzymes; that is, while *gross* photosynthesis is still operating at top speed, the rate of respiration steadily increases, diminishing the net photosynthetic yield.

In the C_4 plants the optimum for F_n lies at temperatures above 30° C—in some cases it reaches 50° C, very high for a biological process (Fig. 35, Table 6). The C_4 pathway for carbon assimilation therefore represents the genotypic prerequisite for colonization of extremely hot habitats. Among the C_3 plants the optimum (like the other cardinal points) is not so much a species character as an adaptation to the

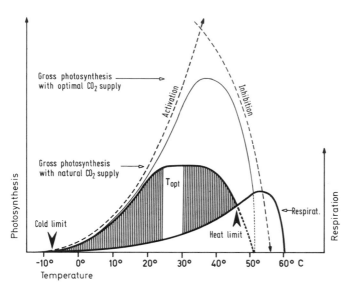

Fig. 34. Diagram of the temperature dependence of photosynthesis and respiration. Gross photosynthesis increases as a result of thermal activation of the enzymes involved, until inadequate CO_2 supply and inhibitory effects (such as enzyme imbalance) lead to a decline in photosynthetic activity. Respiration increases exponentially over an extensive region and at high temperatures, which also damage the protoplasm, falls off rapidly (cf. also Fig. 128). The difference between gross photosynthesis and respiration gives the net photosynthesis (cross-hatched area); note the corresponding positions of the cold limit, the temperature optimum (T_{opt}), and the heat limit for net photosynthesis

thermal conditions in the natural habitat of the plant at the time when growth is proceeding actively. Sciophytes, which only occasionally encounter direct radiation and are warmed less than the plants in sunny locations, function optimally between 10° and 20° C; so too do spring-blooming and high alpine plants, which grow during a season or in a locality characterized by low average air temperature. Herbs in sunny habitats and trees of warm climates, on the other hand, achieve their highest photosynthetic productivity between 20° and 30° C. It will come as no surprise that the lichens, which occupy the extreme outposts of plant life both in the high mountains and in the polar regions, are adapted to the cold climate of the places where they grow. But foliose and crustaceous lichens from warmer countries, and even those from hot deserts, also have optima at lower temperatures. This adaptation is also ecologically reasonable, for really productive assimilation occurs in these lichens only when they are well supplied with water—when they are wet with rain, dew or fog and when the humidity is high. These conditions normally prevail only when the sky is overcast or in the early morning.

The Low Temperature Limit for Net Photosynthesis. Tropical plants function productively only at temperatures of 5—7° C or higher, whereas plants of the temperate zones and cold regions assimilate CO_2 even at temperatures below 0° C. In the higher plants CO_2 uptake is blocked as soon as the assimilation organs begin to

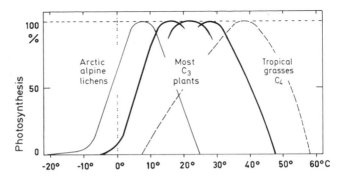

Fig. 35. Temperature dependence and position of the cardinal points for net photosynthesis in different types of plants under light saturation. Lichen data, from Lange (1965), Kallio and Heinonen (1971); C_3 plants, from the data of various authors as cited by Larcher (1969a); C_4 grasses, from Ludlow and Wilson (1971a, b, c)

Table 6. Temperature dependence of net photosynthesis during the growing season, under conditions of natural CO_2 availability and light saturation. (From measurements of numerous authors)

Plant group	Low temperature limit for CO_2 uptake	Temperature optimum of F_n	High temperature limit for CO_2 uptake
1. Herbaceous plants			
C_4 plants of hot habitats	5–7° C	35–45° C (50)	50–60° C
Agricultural C_3 plants	−2 to over 0	20–30 (40)	40–50
Heliophytes	−2 to 0	20–30	40–50
Sciophytes	−2 to 0	10–20	ca. 40
Spring-flowering and alpine plants	−7 to −2	10–20	30–40
2. Woody plants			
Evergreen foliage trees of the tropics and subtropics	0 to 5	25–30	45–50
Sclerophyllous trees and shrubs from dry regions	−5 to −1	15–35	42–55
Deciduous foliage trees of the temperature zone	−3 to −1	15–25	40–45
Evergreen conifers	−5 to −3	10–25	35–42
Dwarf shrubs of heath and tundra	ca. −3	15–25	40–45
3. Lichens			
of cold regions	−25 to −10	5–15	20–30

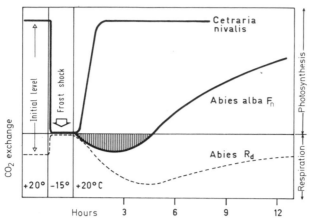

Fig. 36. Net photosynthesis and respiration following a period of frost. The net photosynthesis of the lichen *Cetraria nivalis* recovers completely after a frost shock, within an hour at 20° C. In the fir, photosynthesis is so strongly inhibited after thawing that in light (10 kLx) CO_2 is released for several hours (cross-hatched area). Even afterward, recovery proceeds very slowly and is not always complete. Respiration in the dark by the fir (dashed curve, increasing downward) is enormously heightened for hours after frost shock; the return to normal can take one to several days. (After Lange, 1962; Kallio and Heinonen, 1971; Pisek and Kemnitzer, 1968; Bauer *et al.*, 1969; and M. Huter, unpublished results)

freeze, which occurs in the spring at −2° C and in the winter, in the case of evergreens, at −5° C to −8° C. Many lichens behave differently from vascular plants; they take up and incorporate CO_2 at −10° C and even at −25° C, i.e. even when the thalli are frozen.

The High Temperature Limit for Net Photosynthesis. With very high temperatures, the rate of photosynthesis falls off sharply, and at the same time the intensified rate of respiration frees larger amounts of CO_2. The "temperature maximum for net photosynthesis" is the highest temperature at which all the CO_2 given off by respiration is reassimilated; if the temperature rises further, CO_2 begins to escape. One can therefore consider the high temperature limit for net photosynthesis as a compensation point (the heat compensation point), reached sooner the more heat-sensitive the photosynthesis and the more rapid the enhancement of respiration. Plants of high photosynthetic performance and those with slow respiration have an advantage; in some tropical grasses and in Chenopodiaceae of extremely hot habitats, heat compensation points have been measured at 58° and 60° C—just under the temperature causing death of the leaves.

Aftereffects of Frost and Heat

As long as the temperature is extremely low or high, gas exchange is entirely suppressed. Subsequently, as more favorable conditions reappear, it is only in rare cases that the plants recover immediately.

After freezing, some (but by no means all) lichens resume assimilation without delay, but higher plants cannot do this (Fig. 36). In the latter, cold shock is followed by a

50

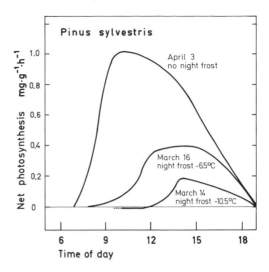

Fig. 37. Aftereffect of night frost on the time course and yield of net photosynthesis of pine twigs during the following day. (After Polster and Fuchs, 1963)

temporary, rapid increase in the rate of respiration. Only after several hours does respiratory activity return to normal. Photosynthesis begins slowly after the plants have thawed out, so that initially the intensified respiration predominates and CO_2 is given off. Only after several or many hours is CO_2 uptake resumed. The lower the temperature and the longer the period of exposure, the more severe and prolonged is the setback. Repeated freezing has the same effect as more severe cold. The daily rhythm of *net photosynthesis after night frosts* is distinguished by a slower increase and lower peak value of CO_2 uptake, the lower the temperature in the preceding night (Fig. 37). A series of night frosts progressively restricts the period during the day that can be used for CO_2 uptake, and thus diminishes considerably the CO_2 uptake of the plants.

Heat, too, impedes gas exchange not only while the temperature is high but afterward as well, so that the original rate of photosynthesis is not restored for days. Particularly heat-sensitive photosynthetic systems are found in algae from the uniformly cool deep-sea water and in some lichens in a turgescent state. In higher plants one must expect impairment of photosynthesis to last for hours or days when the heat compensation point of net photosynthesis has been exceeded.

Temperature Adaptations

The rates of photosynthesis and respiration adapt to the temperature prevailing at a given time.

In the cold season of the year, evergreen woody plants in maritime climatic areas, with winter temperatures rarely below freezing, show a displacement toward lower temperatures of both the temperature optimum and the maximum for net photosynthesis, and sometimes of the minimum as well (Fig. 38). At the same time the level of respiratory activity is readjusted so that respiration still proceeds rapidly enough at the lower temperatures. This adjustment can be brought about by alterations in the relative concentrations of the various enzymes, as well as by the replacement of

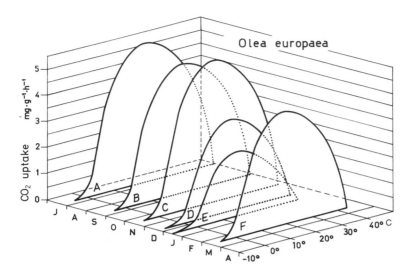

Fig. 38. Temperature dependence of net photosynthetic capacity of the olive tree in the course of the year. *A* New twigs after development is completed; *B* twigs at the end of a dry period in September; *C* during the autumn rainy season; *D* in winter before the first frost; *E* in the winter during a period of night frosts; *F* last year's twigs in the spring. (After Larcher, 1969b)

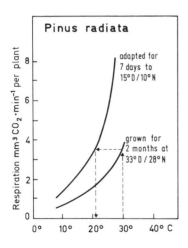

Fig. 39. Adjustment of the temperature-dependence of respiration of pine seedlings to two different temperatures. If one brings warm-adapted seedlings (raised at 33° C by day and 28° C by night) into a cool room (15° C by day and 10° C by night), within a week activity is doubled. Furthermore, the temperature curve of respiration rises more steeply. As a result the plants now respire just as rapidly at 21° C as formerly at 30° C. If the seedlings are subsequently returned to the warm room, the respiratory curve gradually resumes its original shape (not shown in the figure). (After Rook, 1969)

enzymes originally present by isoenzymes having the same action but a different temperature dependence. In this way a state of homeostasis is attained such that respiration can produce about as much ATP as before even though the temperature has fallen by 10° C. Homeostasis of this kind is widespread among poikilothermic organisms, including animals. In plants a compensatory temperature adaptation of respiration can be demonstrated even after a few days, and in alpine plants within a single day. Fig. 39 illustrates and explains this process.

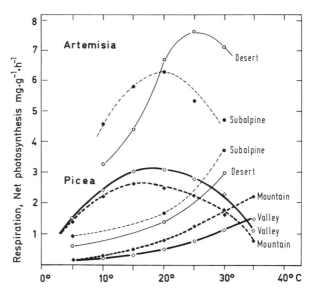

Fig. 40. Influence of adaptation to the habitat upon the temperature dependence of net photosynthesis (peaked curves) and respiration (exponential curves) of *Artemisia tridentata* and *Picea abies*. The desert samples of *Artemisia* were gathered at 1400 m above sea level (mean maximum temperature in the locality in the time preceding the measurement: 38.9° C, mean minimum temperature, 11.7° C); the subalpine samples were collected at 3090 m (mean maximum 19.5°C, mean minimum −2.4°C). The mountain spruce came from 1840 m (mean maximum in July 13.1° C, mean minimum 6.2° C), the valley samples from 600 m (mean maximum 24.0° C, mean minimum 13.0° C). The plants adapted to a cooler climate (dashed lines) display a shift of the temperature optimum for net photosynthesis toward lower temperatures, and steeper slopes of the respiration curves. (After Pisek and Winkler, 1959; Mooney and West, 1964)

In a similar manner, photosynthesis and respiration can adapt to the average temperature in the habitat of a plant. Fig. 40 contains examples of such adaptations. The temperature optimum for net photosynthesis is distinctly lower in plants from cool mountain habitats than in plants from warmer valley locations, and the respiration of the plants adapted to warmer conditions rises with increasing temperature less steeply than that of the mountain plants.

A similar adjustment of respiration to temperature is apparent in the comparison of plants from different climatic regions. Arctic plants at 20° C respire about twice as rapidly as comparable plants of the temperate zone, while tropical plants respire at only half the rate. Warmth accelerates basal metabolism less in tropical plants than in plants adapted to a cool climate.

CO_2 Exchange and Water Supply

Like carbon dioxide, water is used in the photosynthetic process, but it is not in this respect that water shortage can be a limiting factor; more important is the water necessary to maintain a high turgidity in the protoplasm. The metabolic processes of the cell are critically dependent upon water in this sense. In particular, water content influences CO_2 exchange by the turgor-regulated stomatal pore size.

CO₂ Exchange and Degree of Hydration of Thallophytes

In thallophytes the degree of hydration of the cells is matched to the humidity of the surroundings (cf. p. 000). These primitive plants rapidly soak up water when they are sprinkled, but they lose it again quickly through evaporation; thus their water content fluctuates over short intervals with the meteorological conditions and stays in equilibrium with the water-vapor content of the air. As desiccation increases, photosynthetic activity is gradually extinguished, and respiration too is suppressed.

An ecologically important measure is the *humidity compensation point*. The minimum atmospheric humidity for net photosynthesis is about 70% relative humidity (*RH*) for aerial algae, around 80% *RH* for lichens, and for those mosses that can extract water from the air, usually above 90% *RH*.

The photosynthetic apparatus of the thallophytes is well suited to the frequent and pronounced fluctuations in the cellular water content. Completely dry thalli reactivate the photosynthetic process within minutes after they receive water again, even if they have been dried out for a long time. For lichens in dry habitats, this ability enables their very existence, e.g. desert lichens can utilize to the utmost the short period available between their imbibing water at night and drying out again in the morning. The course of CO_2 exchange during the day in a desert lichen, described in Fig. 41, illustrates this behavior under the conditions in the natural habitat.

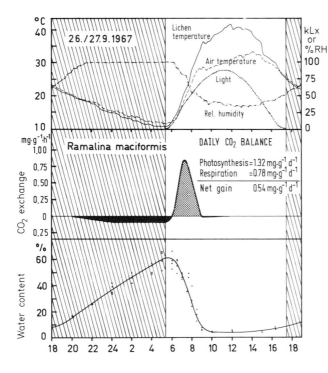

Fig. 41. CO_2 exchange and water content of the desert lichen *Ramalina maciformis* during the course of the day. The nighttime dew supplies the lichen with moisture, but in the morning it rapidly dries out again; cf. also Fig. 84. Stippled area: CO_2 uptake. Black areas: CO_2 release. Cross-hatched: nighttime. (After Lange *et al.*, 1970)

Throughout the night the thalli soak up moisture from the air, and in the early hours of the morning they obtain dew as well. After sunrise only three hours remain for the lichens to fixate carbon; then they become dry and stiff again until nightfall.

Gas Exchange during Water Stress in Vascular Plants

The first effect of water deficiency upon vascular plants is on the stomata, the narrowing of which slows down CO_2 exchange. With increasing desiccation there is reduced hydration of the protoplasm in general, and thus reduced photosynthetic capacity. Normally CO_2 uptake is high only over a narrow range of the adequate water supply level; beyond this it begins to decline and eventually is entirely suspended (Fig. 42). There are therefore two critical points in the curve of gas exchange *vs.* water loss: the point of transition from full performance to the *limited region* and the *null point for gas exchange.*

The first critical point comes at a level of water stress in which the stomata begin to close, causing the stomatal diffusion resistance to predominate over the diffusion resistances inside the leaf (cf. also Fig. 20). If water is supplied after this first critical point has been passed, recovery is rapid.

The second critical point is determined by marked or complete closing of the stomata as well as by the direct effect of water shortage on the protoplasm. Appreciable CO_2 uptake is no longer possible, though the CO_2 freed by respiration can be bound again. Once this state has been reached, a renewed water supply does not lead to an immediate recovery of photosynthesis. Recovery is delayed, and after severe desiccation the original photosynthetic capacity may, under certain conditions, never be achieved again (Fig. 43).

The sensitivity of CO_2 exchange to lack of water, and the positions of the foregoing two critical values, are to a large extent characteristic of a plant species, but they are also adaptable. The limiting values for a number of species representing different ecophysiologic types are shown in Fig. 42. It is apparent that sciophytes and leaves

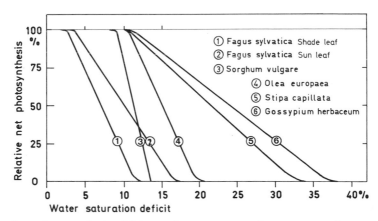

Fig. 42. Restriction of net photosynthesis with increasing water stress. As a measure of water deficiency, the water saturation deficit of the leaves (see p. 158) is indicated. (After Pisek and Winkler, 1956; Larcher, 1963a; El-Sharkawy and Hesketh, 1964; Gloser, 1967)

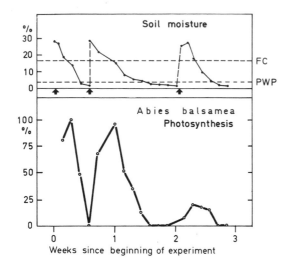

Fig. 43. Effect of drought on the net photosynthesis of year-old seedlings of the balsam fir. As a result of watering (arrows), the soil moisture was raised above field capacity (*FC*, see p. 144. Thereafter the soil dried out to below the permanent wilting percentage (*PWP*, cf. p. 145). After a brief period of drought, net photosynthesis recovered quickly and completely, but after the soil moisture had remained below the *PWP* for several days recovery was incomplete. (After Clark, 1961)

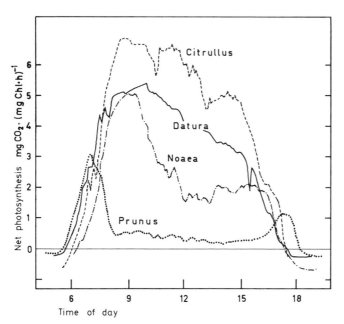

Fig. 44. Daily fluctuation in net photosynthesis of plants differing in sensitivity to drought and in the degree of drought stress to which they are exposed, at the end of the dry period in the Negev Desert. *Citrullus colocynthis* and *Datura metel* were watered, and the F_n values were reduced somewhat only during the hottest hours of the day and in the afternoon. The desert plant *Noaea mucronata*, after extensive CO_2 uptake in the morning, toward noon vigorously restricts its photosynthesis. *Prunus armeniaca* suffers considerably from lack of water toward the end of the dry season, and appreciable CO_2 uptake is possible only in the early morning and late evening. (After Schulze et al., 1972)

adapted to the shade have a very sensitive reaction to slight losses of water; millet, a C_4 plant with especially effective stomatal regulation, behaves similarly. Heliophytes, on the other hand, tolerate larger water losses before they cut off gas exchange.

The closing of the stomata when the water supply is impaired is primarily a water-conservation measure. The different forms this behavior can take are therefore most readily understandable from the point of view of water balance in the plant, which is also true of the daily rhythm of net photosynthesis in natural surroundings. In Fig. 44, typical *diurnal fluctuations in net photosynthesis with suboptimal water supply* are summarized. The principle exemplified is the following: the more sensitive a species is to lack of water and the dryer the conditions, the earlier in the day restrictions are imposed upon assimilative activity. The succulents with crassulacean acid metabolism, finally, keep their stomata closed throughout the day in dry periods; photosynthesis can proceed nonetheless, since it is furnished with CO_2 by the breakdown of malate stored during the night.

Gas Exchange and Soil Factors

Photosynthesis and respiration are influenced by the ground temperature, by the mobility of water and the air content in the soil, and by mineral nutrients.

The temperature of the soil, its oxygen concentration, and the amount of water in its interstices act first upon the rate of respiration in the roots, their growth, and their absorptive activity. Indirectly, however, the whole plant is affected. This fact is clearly evident in cases of flooding and stagnant groundwater, which only a few vascular plants can survive.

The influence of the *nutrient salt supply* upon photosynthesis and respiration is extremely varied. In soils not seriously deficient in particular nutrients, the availability of minerals is less critical than the climatic factors. Nevertheless, it is almost always possible to enhance the yield of photosynthesis by the artificial provision of nutrients. Mineral nutrients can influence carbon metabolism both directly and indirectly *via* the synthesis of new tissue and growth. Direct effects upon photosynthesis and respiration result from the fact that the minerals either are incorporated in coenzymes and pigments or participate directly as activators in the process of photosynthesis (cf. Tables 17 and 18). Manganese, for example, acts as an activator of photolysis, and potassium is involved in the electron-transport system on the thylacoids. Nitrogen and magnesium are components of chlorophyll; various enzymes include iron, cobalt and copper, and phosphate is a component of nucleotides.

The lack of minerals, as well as alterations in relative amounts of the elements taken up, can affect the chlorophyll content and the number, size and ultrastructure of the chloroplasts; this applies even if the elements in question, e.g. iron, are not themselves incorporated into the chlorophyll molecule. In conditions of nitrogen and iron deficiency, chloroses are observed, which cause a diminution of CO_2 uptake to less than $1/3$ (Fig. 45). Lack of magnesium can have similar consequences. The chief result of insufficient chlorophyll is that the plants cannot make full use of intense light—they behave like sciophytes.

Mineral nutrients further affect gas exchange by influencing the behavior of the stomata, and by their effect on other properties of the leaves such as their anatomic

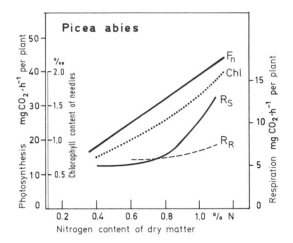

Fig. 45. Influence of nitrogen supply on net photosynthesis (F_n), respiration in the shoots (R_S), and respiration in the roots (R_R) of young spruce plants. Photosynthetic capacity changes in proportion to the chlorophyll content (*Chl*) of the needles. (After Th. Keller, 1971a)

structure, size, life span, and above all their number. Under nitrogen deficiency small leaves develop, with stomata that are less movable, whereas too much nitrogen causes excessive respiration and thus reduces the photosynthetic yield.

The Interplay of External Factors Affecting CO_2 Exchange

Environmental factors do not, of course, act in isolation. The gas exchange rate of plants is an expression of the interplay of many internal and external environmental factors, the individual roles of which are not easy to unravel. Of these factors, one is

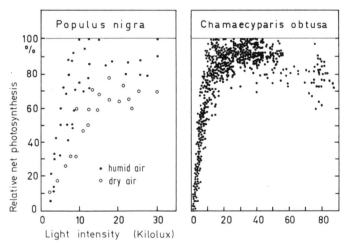

Fig. 46. Light-dependence of net photosynthesis in the open under different meteorological conditions. *Left graph*: The circles are data from mature poplars on clear days with low humidity; the dots are data taken on cloudy to slightly overcast days with high humidity. (After Polster and Neuwirth, 1958). *Right graph*: Data from pot-grown seedlings during summer (adjusted to a uniform 20° C to eliminate the influence of temperature). (After Negisi, 1966)

usually rate limiting at any given time. For example, with increasing illumination the optimum and maximum temperatures for net photosynthesis are shifted toward higher values. This effect is advantageous for the plant, since intense irradiation is always associated with warming. If the strong radiation leads to overheating (or an intolerable loss of water), then the light loses its role as the limiting factor and CO_2 uptake declines. As a result, in the field one does not obtain saturating curves like those measured in laboratory experiments (which attempt to treat one factor at a

Table 7. Reduction of CO_2 uptake (the average measured F_n as % of the maximum possible) in the natural habitat, by various inhibitory factors

A. Year-old seedlings of *Pinus densiflora* and *Cryptomeria japonica* in Japan. (From Negisi, 1966)

Factor	Average annual reduction		Reduction during main growing season (April to September)		Period of greatest effectiveness
	Pinus	*Crypto-meria*	*Pinus*	*Crypto-meria*	
Reduced photosynthetic capacity	−36%	−25%	− 8%	−10%	Winter
Lack of light at twilight and when the sun's elevation is low	−16%	−14%	− 5%	− 5%	Winter
Lack of light due to clouds	−11%	− 9%	−17%	−10%	Early summer
Temperature too low or too high	− 7%	−10%	−15%	−25%	Spring (cold) Midsummer (heat)
Total reduction	**−70%**	**−58%**	**−37%**	**−40%**	
Residual CO_2 uptake	*+30%*	*+42%*	*+53%*	*+50%*	

B. Mature beech in northwest Germany. (From E. D. Schulze, 1970)

Factor	Average reduction during growing season	Maximal reduction during periods of unfavorable weather
Lack of light at twilight	−22%	
Lack of light due to clouds and fog	−16%	−56%
Unfavorable temperature	− 3%	−15%
Dryness of air	− 2%	−13%
Total reduction	**−43%**	
Residual CO_2 uptake	−57%	

time), but rather curves with maxima (Fig. 46). In the range of optimum illumination in the field, light is sufficient to stimulate photosynthesis maximally but not so great that the disadvantageous side effects of strong radiation are noticeable. In this natural interplay the external factors are only rarely and briefly found to be optimal in themselves and yet so related to one another that maximum photosynthesis is achieved. On the average, the maximum daily values for CO_2 uptake reach only 70—80% of the actual photosynthetic capacity. This is true of herbaceous plants as well as for trees.

Under the climatic conditions prevailing at intermediate latitudes, lack of *light* when the sun is low and obscured by clouds is the foremost factor limiting the yield of CO_2 assimilation by plants in the field (Table 7), especially for those in dense stands. Unfavorable temperature has little effect on deciduous trees in the temperate zone, and such influence as it has is most apparent in the spring; on the other hand, evergreen trees in cold-winter areas suffer significant setbacks from cold, and the vegetation of warmer lands, from heat. From a global point of view, lack of *water* is the most significant factor limiting assimilation. Though the amount by which CO_2 uptake is diminished during days when water is limiting has been measured, there has been no quantitative analysis of the long-term effect in plants of dry regions. From the diurnal fluctuations in CO_2 exchange it can be inferred that evergreen shrubs of maquis and bushland achieve daily photosynthetic maxima in the dry periods only $^1/_3$ to $^4/_5$ of those in the rainy season. These numbers are not particularly informative, since they vary greatly as a function of local ground-water reserves; moreover, they become still lower under prolonged drought.

Carbon Budget and Dry-Matter Production in Plants

The Gas Exchange Balance

Where the synthesis of organic dry matter and the ability of a species to compete effectively in its habitat are concerned, the determining factor is not so much the brief peak values of photosynthesis, but rather the average yield of CO_2 uptake. The average values for net photosynthesis even under favorable climatic conditions, like those prevailing in the temperate zone, amount to 50—60% of the daily maxima (i.e., 30—50% of the actual photosynthetic capacity).

The Time Factor

The decisive factor in carbon fixation is the time span over which a high rate of CO_2 acquisition is possible. This comprises the hours of daylight during the leaf-bearing time of year, insofar as assimilation is not blocked by frost, heat or drought. From the sum total of daily carbon intake, the nightly CO_2 release from the leaves must first be subtracted. The resulting net consumption of CO_2 by photosynthesis in the course of 24 hrs is the *daily balance*, and the sum of the daily balances gives the *annual balance* of CO_2 exchange. The daily balance is positive if the intake during the day exceeds the loss at night, and it is greater, the more favorable the constellation of factors influencing photosynthesis during the day, and the shorter and cooler the night. When CO_2 uptake has been possible only briefly or inefficiently, there is little surplus to be assimilated. This is a routine occurrence during dry periods and in dry habitats (cf. Figs. 41 and 44), as well as after frost. In deep shade only slightly positive and—especially when the nights are warm—sometimes negative daily balances are common and in the extreme case can result in "starvation" of the plants. This is clearly one of the factors determining plant distribution.

Green and Non-Green Components of the Plant Mass

Plants consist not only of green, i.e. photosynthetically productive, tissues, but also of others that simply respire and must be nourished by the leaves. In the overall CO_2 budget, therefore, the respiration of all non-green tissues must also be taken into account. The situation is of course most apparent in cases where the axes of shoots, the roots, the flowers and the fruits of a plant make up a large proportion of its mass as compared with the foliage (cf. Table 8).

The Overall CO_2 Balance

The CO_2 balance of an entire plant involves the total gross photosynthesis ($W_L \Sigma_l F_g$, where l is the number of daylight hours in a year and W_L is the weight of the leaves)

Table 8. The proportion of total mass (dry matter) of plants accounted for by assimilation organs, axial structures and roots. (Compiled from the original data of numerous authors)

Plant	Green mass (photo-synthetically active organs)	Purely respiratory organs	
		Woody stems above ground	Roots and subterranean shoots
Evergreen trees of tropical and subtropical forests	ca. 2%	80—90%[a]	10—20%[a]
Deciduous trees of the temperate zone	1—2%	ca. 80%[a]	ca. 20%[a]
Evergreen conifers of the taiga and in mountain forests	4—5%	ca. 75%[a]	ca. 20%[a]
Alpine scrubwood	ca. 25%	ca. 30%[a]	ca. 45%[a]
Young conifers	50—60%	40—50%[a]	ca. 10%[a]
Dwarf ericaceous shrubs of heath and tundra	10—20%	ca. 20%[a]	60—70%[a]
Meadow plants	ca. 50%		ca. 50%
Alpine grassland plants	ca. 30%		ca. 70%
Steppe plants			
wet years	ca. 30%		ca. 70%
dry years	ca. 10%		ca. 90%
Plants of the high mountains	10—20%		80—90%

[a] The greater part of the mass is dead supporting structures.

and the total annual respiration of leaves ($W_L \Sigma R_L$), shoot axes, flowers and fruits ($W_S \Sigma R_S$), and roots ($W_R \Sigma R_R$). That is,

$$CO_2 \text{ Balance} = W_L \Sigma_t F_g - W_L \Sigma R_L - W_S \Sigma R_S - W_R \Sigma R_R \qquad (12)$$

To set up a complete gas-exchange balance for a plant is a laborious undertaking. It involves separation of the CO_2 exchange of the part of the plant above ground into divisions (top of the plant, base of the crown, stems) and measurement of root

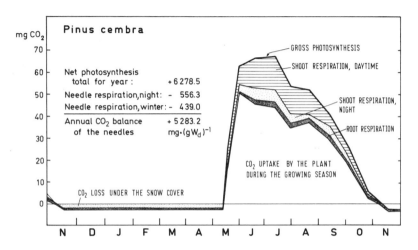

Fig. 47. Variation in the daily CO_2 balance of young stone-pines at the alpine tree line over a year. Part of the CO_2 gained by photosynthesis is lost the same day due to the respiration of the shoots and roots. In winter the daily balance is usually negative or at best zero; the CO_2 loss during the 6 winter months is subtracted from the CO_2 acquired during the growing season, to obtain the annual balance. (After Tranquillini, 1959)

respiration by day and night over the whole year; in addition, the proportions by mass of the different organs are determined. It goes without saying that the external factors effective at the site (for example, light distribution, leaf temperatures, the humidity of the air near the leaves, temperature fluctuations in the vicinity of the stem and roots, availability of water, etc.) must also be noted if the carbon balance is to be meaningful. Fig. 47 shows an example of a yearly balance of CO_2 turnover, a typical one for young trees.

The units of gas-exchange balance are g or kg CO_2 per plant, per day or per year. Conversion factors can be used to express this quantity in terms of organic dry matter ar carbon content (see p. XIV). Thus one can convert *from gas Exchange to production of matter* by the plant; the latter, of course, requires a positive gas-exchange balance.

Dry-Matter Production

Assimilated carbon not lost by respiration (i.e., the surplus in the CO_2 budget) increases the dry matter of a plant and can be used for growth and for laying down reserves.

Growth Analysis

The accumulation of carbon is evident in increased weight and can thus be determined directly by weighing the gathered and dried plants. This method is used to measure growth in studies of the influence of environmental factors on carbon metabolism. As a measure of dry-matter production, the net assimilation rate (NAR) is computed as follows:

$$NAR = \frac{dW}{dt} \cdot \frac{1}{A} \qquad (13)$$

In this equation, formulated by F. G. Gregory, dW is the increase in mass of dry matter in a given time interval (dt). This rate of growth is referred to the assimilation area (A), the photosynthetic activity of which is responsible for the production of matter. The NAR is given in g dry matter (or g carbon) per dm^2 leaf area of the plant and per day or week. Formula 13 defines NAR at any instant. To estimate NAR from a set of measurements of A and W at different times, various procedures are available. One of these is given by the formula of D. J. Watson,

$$NAR = \frac{W_2 - W_1}{A_2 - A_1} \cdot \frac{\ln(A_2/A_1)}{t_2 - t_1} \qquad (14)$$

where W_1 and A_1 are the dry weight and leaf area, respectively, of the whole plant at time t_1

W_2 and A_2 are the dry weight and leaf area, respectively, at the later time t_2.

With herbaceous plants a time interval ($t_2 - t_1$) of 1–2 weeks is usually chosen.
This equation results from calculating the average value of NAR over the time interval $t_2 - t_1$, under the assumption that both A and W change linearly with time. The logarithmic function ("ln", natural log) arises when the associated integral is evaluated.
The "dry weight" required here is that of the organic matter of the plant; it is therefore not accurate to use simply the value determined by weighing the dry plant, which includes not only carbon compounds but also minerals (these average 3–10% of the total dry weight). One must subtract from the dry weight of the whole sample the weight of the ashes after the sample is burnt. The average net assimilation rates of temperate-zone herbaceous plants are in the range 0.05–0.1 g dry matter per dm^2 leaf area per day; woody plants (young trees, dwarf shrubs) assimilate at average rates of 0.01–0.02 $g \cdot dm^{-2} \cdot d^{-1}$. The highest rates achieved by herbaceous crops for brief periods during the growing season lie between 0.12 and 0.25 $g \cdot dm^{-2} \cdot d^{-1}$.

Utilization of Photosynthates and the Rate of Growth

Plants consist largely of carbohydrates; carbohydrates provide the material for construction of the cell walls, and comprise 60% or more of the dry matter of higher plants. The carbohydrates produced in CO_2 assimilation must be distributed throughout the plant in a systematic way; distribution is controlled by demand (for energy, growth or differentiation) and by coordinating mechanisms, some of which involve hormones. There are a number of characteristic ways in which plants budget their photosynthates, depending upon their level of organization and life form. These differences are evident in the substances produced and the rate of growth.

Type 1: Planktonic Algae

Planktonic algae exist as single cells or form colonies or simple groups of cells. The typical algal cell supplies only itself with carbon and need not produce any surplus for other cells. Within the cell there is a favorable ratio between the sites of production and cell components that consume photosynthates: in *Chlorella* the chromatophores take up about half of the volume of protoplasm. This being the case, it is not surprising that algal cells well supplied with nutrient elements and light accumulate large surpluses and grow rapidly. They soon reach their terminal size and then proceed to divide. The autotrophic single-celled organism employs its yield of synthesized materials to increase the number of individuals, i.e. for reproductive processes. There is a direct relationship between photosynthetic yield and the number of divisions per day. The rate of growth in phytoplankton is thus usefully expressed as the increase in population density or the number of divisions per unit time.

Type 2: Annual Plants

These are frequently rather small herbs that must make best use of a short period of time in which conditions are favorable for growth, flowering and the setting of fruit. They are found primarily in dry regions, where they complete their life cycles in a few weeks to months. These plants must employ their photosynthates in such a way that an abundance of tissue is formed in the shortest possible time. Annuals do this even when a rather long time is available for growth. Summer grain, sunflowers and other annual crops thus yield particularly large harvests.

The operating principle of all annual plants consists in first using the greater proportion of the photosynthates for the formation of leaves, which then participate in production and increase the intake of the plant. While these photosynthetically active organs are being developed preferentially, the mass of those parts that only respire remains small, which improves the overall balance. In the flowering phase, the distribution system switches to favor the reproductive organs, which receive such a large share that all other parts of the plant are supplied with little more than needed to maintain themselves—the older leaves even shrivel up. Accordingly, in the course of the life cycle the proportions of leaves, axial structures, roots, and reproductive organs in a plant change considerably (Fig. 48). The greatest change is in the fraction of the overall mass represented by leaves, which sinks from 30—60% during the elongation phase to 10—20% by the time the fruits are ripe. At this time, in the

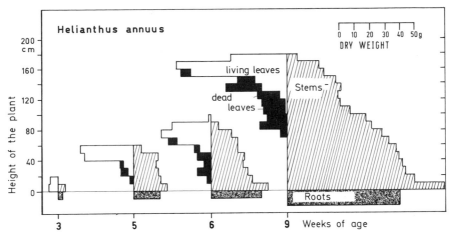

Fig. 48. Distribution of dry matter in growing sunflowers among photosynthetically active foliage (on the left in each case), axial material (cross-hatched), and roots (stippled). The mass of dead leaves is shown in black. (After Hiroi and Monsi, 1966)

sunflower, 90% of the photosynthates produced daily moves into the fruits (Fig. 49 left).

Under environmental conditions conducive to plant life, this way of investing assimilation products selectively guarantees both luxuriant growth and lavish fruiting. When local conditions are less favorable, on the other hand, particularly when there is a shortage of water or when the soil is poor in nutrients, the plant is forced to build up an extensive system of roots; this is done at the price of leaf-area development and leads to a smaller photosynthetic yield as well as deterioration of competitive ability (Table 8). Annual plants are primarily adapted to making use of advantageous—though short-lasting—situations, and are less able than other plants to endure prolonged unfavorable conditions.

Type 3: Perennial Herbs

The herbaceous plants which live for several years usually at first undergo a development similar to that of the annuals. But after their vegetative structures are formed, they lay down reserve supplies before proceeding to bloom. Toward the end of the first growing season the excess photosynthates are diverted to the stems and above all into the subterranean parts of the plant, which may develop into massive storage organs. Flowers are formed only after the plant has accumulated sufficient capital to draw upon for that purpose. Photosynthates stored in one year are first used in the next to elaborate the shoot system. The synthesis of new substance builds up rapidly and thanks to the availability of stored material is largely independent of the factors affecting productivity in the spring. Once the plant is ready to flower, if the food supply is adequate, the flowers and fruits take precedence over the storage processes. Afterward, near the end of the season, photosynthates move preferentially to the subterranean parts of the plant, which increase correspondingly in weight (Fig. 49 right).

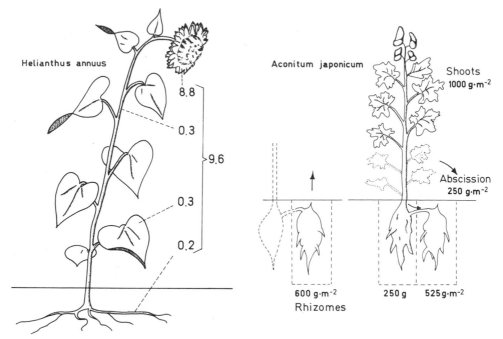

Fig. 49. Distribution of photosynthates in one-year-old and two-year-old plants. *Left: Sunflower in bloom.* The numbers indicate the daily incorporation of carbon (g C · m⁻² · d⁻¹) into leaves, axial structures, flowers and roots of the plants in a stand of sunflowers. (After Eckardt *et al.*, 1971). *Right: Monkshood.* Distribution and shifting of dry matter when the shoots are forming and when the rhizome reserves are being filled up for the next year. (After Iwaki and Midorikawa, 1968)

Perennial plants have the advantage wherever the period of time favorable to production is not long enough to permit sufficient assimilation for flowers and fruits to be formed, as well as in cases where the plants bloom so early that the necessary materials cannot be provided by the available mass of leaves. This applies, for example, to spring geophytes, many of which open their flowers before the leaves have unfolded. Alpine plants must accomplish flowering and ripening of seeds during the short mountain summer; their accumulation of photosynthates is subject to many uncertainties. Finally, these considerations also apply to steppe plants, which must utilize the times between winter cold and summer drought for their life cycle. A trait all these species have in common is the presence of storage organs such as rhizomes, tubers, root thickenings and bulbs. Moreover, these species frequently develop an extensive root system, so that the subterranean dry mass amounts not uncommonly to twice and sometimes even to four times the mass of the parts above ground (Table 8).

Type 4: Trees

The tree—a highly differentiated, large vascular plant—manages its carbon supplies in a way suited to its long lifetime. Even in youth a large fraction of the photosynthates is used for growth of the stem. In the first years of life the leaf mass can make up half of the overall dry substance of the plant, but with increasing size the ratio of leaf mass to stems is altered, the leaf mass growing only slightly while trunk and branches become steadily thicker and heavier. Foliage comprises only 1–5% of the total mass of mature trees (Table 8), and these leaves must thus supply the materials for maintenance and growth to parts of the tree amounting to many times their own weight. The consequence is a modest acquisition of carbon and increase in mass as compared with herbs, but that is no disadvantage in view of the long lifespan. Even after maturity the tree increases its mass of wood from year to year; from the standpoint of the photosynthate budget, this represents inaccessible capital, since it is permanently withdrawn from the metabolism of the tree. Organic matter tied up as wood can be used in metabolism only by other components of the ecosystem; the tree stores such matter not for itself but as a member of a food chain. On the other hand, their growth form necessitates this great expenditure on supporting tissue by woody plants. It procures decisive competitive advantages for the trees over herbaceous plants in areas with long production periods; the herbs are slowly but surely overshadowed by the ever taller woody plants.

In correspondence with their size and differentiated infrastructure, the distribution of photosynthates in trees takes place according to a complicated scheme.

In *deciduous trees*, the carbohydrate stores are emptied shortly before the leaves begin to unfold, the substances being sent to the buds and later to the new shoots. About a third of the reserve materials serves for the building up of assimilation surfaces, which very soon operate with a positive balance and in their turn contribute to the further formation of the leaves and shoots in the new growth. After the foliage is completely formed, it supplies the tree with photosynthates. As a rule flowers and developing fruit are supplied preferentially, next in order is the cambium, and last the newly forming buds and the depots of starch in roots and bark. Floral primordia are formed in numbers depending on the amount of material left. This distribution scheme results in a competition between fruit and growth of supporting tissue, and if the photosynthates are in short supply vegetative but not reproductive buds are formed for the next year. The food supply and the amount of fruit set thus affect the increase in woody tissue and the frequency of flowering of the tree. The cost of reproduction, in terms of photosynthates, is considerable; in pines it amounts to 5–15%, in beeches 20% or more, and in apple trees as much as 35% of the net annual yield from photosynthesis. Many trees in the temperate zones can therefore produce an abundance of fruit only at intervals of several years: broad-leaved trees every 2–3 (in exceptional cases 5) years as a rule, and needle-bearing trees every 2–6 (10) years. Near the polar limits of plant distribution, and in the mountains, the fructification intervals become considerably longer. At the end of the growing season the surplus photosynthates are moved into the woody tissue and the bark of branches, trunk and roots and stored. In the tropics and in dry regions the trees pass through several seasonal storage periods—four in the case of fig trees, for example.

Evergreen woody plants of the temperate zones do not produce new shoots as soon as the winter dormant period is past, for they still have the assimilation organs of the previous year. When buds do begin to open, the carbon taken up by these old organs in the spring can meet a large part of the demand, and the rest comes from reserves in axis and roots. Because of this "head start" the new leaves mature relatively rapidly—even though evergreen leaves as a rule incorporate three times as much dry matter for a given area as does the delicate foliage of deciduous trees. Moreover, sufficient photosynthates remain for cambial growth. Evergreen leaves—and the associated extension of the productive period—become particularly advantageous wherever a long winter, or a summer dry season, restricts the growing season. In the mountains, in the northern forest belt, and in regions where aridity limits tree growth, evergreen woody plants generally gain dominance over deciduous plants.

Distribution and Transport of Photosynthates

The products of CO_2 assimilation are constantly being translocated within a plant—from the leaves and other photosynthetically active tissues (green bark, and parts of flowers such as the awn) to sites where they are consumed or stored, and from the storage depots to growth zones and into seeds and fruits. Fig. 50 illustrates the continual shifting of photosynthates in the course of development of an annual grain plant.

The products are conducted through the lumens of sieve tubes, narrow passages that also present high filtration resistance. Nevertheless, the rate of movement of this stream of photosynthates is considerable, since the sieve-tube sap is usually very concentrated. In trees the sieve-tube sap contains an average of 20—30% dry matter, mainly sucrose, and in herbs somewhat less. The mechanics of long-distance transport of photosynthates are not yet known in detail, but many arguments favor mass flow along a concentration gradient. In addition, there is an active component involving the companion cells.

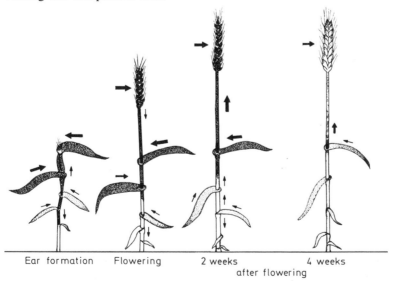

| Ear formation | Flowering | 2 weeks | 4 weeks |
| | | after flowering | |

Fig. 50. Formation and distribution of photosynthates in wheat plants. The darker stippling indicates regions of particularly productive assimilation, and the thickness of the arrows shows the relative rate of transport of the products. (After Stoy, 1966)

Productivity and Carbon Turnover in Plant Communities

The Productivity of Stands of Plants

The Basis of Productivity

The quantity of organic dry matter formed per unit time by the vegetation covering a given area is the measure of the *productivity of a stand of plants*. The units of productivity may be metric tons of organic dry matter per hectare per year, or grams per square meter per day. The production (P) of a plant community is greater, the higher the assimilation rates (NAR) of the plant species composing the community, the more completely the available light is captured by the assimilation surfaces (the leaf-area index, LAI), and the longer the time in which the plants can maintain a positive gas-exchange balance [duration (t) of the production period]. That is,

$$P = NAR \cdot LAI \cdot t \qquad (15)$$

Leaf-Area Index and Productivity

The extent and arrangement of the assimilation surfaces is taken into account by the *LAI*. The degree to which the layers of foliage are superimposed is optimal for production if the PhAR is absorbed as completely as possible during its passage through the canopy of leaves. In stands of cultivated plants this is frequently the case with a *LAI* of about 4 (Fig. 51). If the density of foliage were less, the light available to individual plants, and thus their *NAR*, would be greater, but with respect to the yield per unit ground area an open stand of plants is less productive than a closed stand. If the plants are too closely spaced and the foliage overlaps too extensively, the light in the most shadowy places is no longer sufficient to keep the CO_2 balance positive at all times; thus the yield per unit area will be reduced.

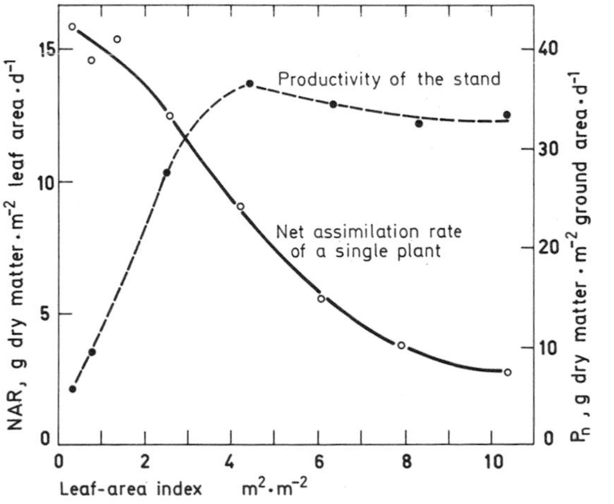

Fig. 51. Relationship between the net assimilation rate of maize plants and the productivity of stands of maize as a function of leaf-area index. (After Williams *et al.* as cited by Baeumer, 1971)

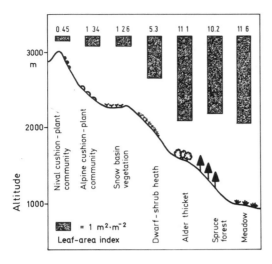

Fig. 52. Leaf-area index of various types of vegetation at a sequence of different altitudes in the Alps. At the tree line, and at the zone of transition to open communities covering a small proportion of the ground, the leaf-area index decreases sharply. (After Vareschi, 1951, extended by data from Larcher et al., 1973)

The density of the foliage of individual plants and the closeness of the plants (that is, the degree of cover) are more than just important factors affecting production—in fact, each is itself affected by production. With an unfavorable food supply and a scarcity of water, the plants lack the raw materials for the synthesis of an extensive leaf system, and the LAI remains insufficient. In dry regions, on poor stony soil, and in areas with a very short production period there arise open plant communities, the LAI of which falls to minimal values as the amount of cover decreases. This can be seen especially clearly if a series of communities at different altitudes is considered (Fig. 52). When the closed plant cover gives way to an open one, the LAI decreases rapidly and irregularly from place to place; in such cases the LAI depends primarily upon the amount of cover and reflects to a smaller extent the degree to which the leaves overlap. In very dense plant communities (for example, certain coniferous forests), an increase of the LAI above 12—14 is prevented by lack of light.

A relationship similar to that between LAI and productivity exists between productivity and amount of chlorophyll per m² of ground. A related measure, used chiefly to characterize the degree of overlapping of photosynthetically active layers in bodies of water, is the chlorophyll content in the plankton-filled column of water below 1 m² of water surface.

Production Period and Yield

Just as the duration of the production period is crucial to the carbon balance of an individual plant, this factor also determines the annual photosynthetic yield in plant communities in a given region. A considerable yield is obtained despite moderate rates of assimilation if the time span favorable to assimilation is long enough—as it is, for example, in warm-temperate, subtropical and tropical humid regions. Where assimilatory activity is possible only during a relatively short period—as in the high mountains, in the arctic, and in dry regions—the yields remain small even if the

Fig. 53. Dependence of net productivity of plant cover upon latitude and the local water supply. The yield of assimilation is reduced at northern latitudes due to lack of light and heat and to the associated shortening of the vegetation period; in more southern latitudes it is reduced primarily by lack of water and an unfavorable distribution of precipitation. (After Bazilevich *et al.*, 1971)

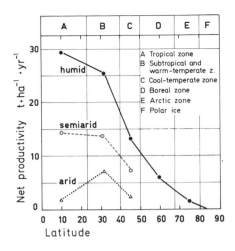

plants have a great photosynthetic capacity or can make particularly efficient use of the photosynthates (Fig. 53).

On continental land masses, the time available for highly productive synthesis decreases in the following sequence, according to the growth form of the plants and the climatic conditions:

1. Evergreen plants of warm, humid regions, which carry on photosynthetic activity throughout the year (for example, in the tropical rain forest)

2. Evergreen plants in which the production period is interrupted
(a) by a cold season (for example, boreal conifers)
(b) by a dry season (for example, shrubs of the Mediterranean maquis)

3. Seasonally green plants (deciduous woody plants and herbs)
(a) which utilize fully the foliated season, in regions with high precipitation (for example, foliage trees in the temperate zone)
(b) which utilize the foliated season only partially because of insufficient light (e.g., undergrowth in deciduous woodland)
(c) which utilize the foliated season only partially because of dryness (e.g., steppe and dry woodlands)

4. Plants which take up carbon during short periods of production between longer unfavorable periods; these include
(a) tracheophytes in deserts with erratic precipitation (100—120 favorable days)
(b) tracheophytes of the arctic and high mountains (60—90 favorable days)
(c) mosses, lichens and aerial algae which take up carbon briefly after they have been wet or when the air humidity is high.

In the cold seas at high latitudes the production period is limited to the polar summer, which permits assimilative activity for several weeks, without interruption at night.

71

Carbon Turnover by Plants as Part of the Ecosystem

The Equation for Net Photosynthetic Productivity

The *carbon balance* of a plant community is determined from the difference between intake and output. Intake is measured as the overall quantity of carbon fixed by photosynthesis in the course of a year. This gross productivity P_g cannot be measured for land plants in the field; therefore (as for gross photosynthesis) a rough estimate is computed from the net productivity and the respiration of the stand.

$$P_g = P_n + R \qquad (16)$$

The net yield of photosynthesis is used for building up organic matter, part of which becomes detritus in the course of the year and is lost (L) or is grazed by consumers (G). These deductions from the net yield include the shedding of leaves, flowers, fruits and dead branches, the decay of dead roots, consumption by animals and parasites, and the release of photosynthates in fluid form *via* the excretions of roots

Table 9. Productivity and loss of organic dry matter in forests (annual balance); all data in metric tons dry matter per hectare per year

Stand	Beech wood, Denmark, 60 years old		Tropical rain forest, Thailand	
Authors	Mar-Möller *et al.* (1954)		Kira *et al.* (1964) Yoda (1967)	
LAI	*5.6*		*11.4*	
Increase in biomass, ΔB		in % P_g		in % P_g
foliage	0		0.03	
stems	5.3		2.9	
roots	1.6		0.2	
Total	**6.9**	*35%*	**3.13**	*2%*
Loss, L				
foliage	2.7		12.0	
stems	1.0		13.3	
roots	0.2		0.2	
Total	**3.9**	*20%*	**25.5**	*20%*
$P_n = \Delta B + L$	**10.8**	*55%*	**28.6**	*22%*
Consumption in respiration				
foliage	4.6		60.1	
stems	3.5		32.9	
roots	0.7		5.9	
Total	**8.8**	*45%*	**98.9**	*78%*
$P_g = P_n + R$	**19.6**	*100%*	**127.5**	*100%*
k_P	**2.23**		**1.29**	

and consumption by symbionts (mycorrhiza). The remaining net yield goes to increase the plant mass per unit area of ground (the biomass B, actually phytomass); it represents the annual change in the dry matter comprising the stand (ΔB).

$$P_n = \Delta B + L + G \tag{17}$$

All these quantities can be determined directly, and their sum is the measure ordinarily used to express net productivity (for an example of such a calculation see Table 9). The measurements are not simple to make in natural stands of plants; data for production by forests and other perennial, many-layered plant communities should be considered as guidelines only, unless they are confirmed by different procedures, applied at the same time.

The productivity equation is ascribable to P. Boysen-Jensen, who as early as 1932 clearly distinguished these relationships and thereby initiated the analysis of causative factors in the field of production ecology. Boysen-Jensen called the increase in biomass (positive ΔB) "net production"; the currently accepted denotation of net primary productivity P_n in ecological publications is the total productivity of dry matter by plants. Equation (17) reflects this customary international usage by production ecologists.

The Proportion of Losses Due to Respiration

A considerable fraction of the amount of carbon obtained by photosynthesis is respired and thus is unavailable for incorporation into the tissue of the plant. The "operating cost" of respiration, which must be subtracted from the gross production, can be expressed by the productivity coefficient k_P, the ratio of gross productivity (P_g) and total respiration (R) of a stand.

$$k_P = \frac{P_g}{R} \doteq \frac{P_n + R}{R} \tag{18}$$

k_P has the value 2 if the same amount of photosynthate is metabolized in respiration as is retained as dry matter. As a dimensionless quantity, k_P may be expressed in the form of a percentage or a fraction (as in Table 9). A coefficient of 2 corresponds to use of 50% (or 0.5) of the photosynthates for respiration.

In populations of plankton, fields of grain and meadows the coefficient is 10 or a little less; that is, it is similar to the coefficient for the photosynthetic capacity of leaves (k_F; cf. Formula 10). In stands of woody plants arranged in several strata (forests, dwarf-shrub heaths), with a relatively large amount of photosynthetically unproductive mass, it falls to 1.5–3, or about one-third of k_F. Forests in the humid tropics respire especially rapidly, and use for the purpose more than 70% of the photosynthetic uptake (Table 9).

Loss as Detritus and by Grazing, and Its Effect on Biomass

The fraction of the annual net primary production represented by the losses L and G is a critical factor in the carbon balance of an ecosystem. Depending on the yield of net primary production and the amount lost, the organic mass composing the community may either increase (ΔB positive), stay the same ($\Delta B = 0$), or decrease (ΔB negative). Which of these possibilities is realized in an ecosystem depends chiefly upon its species composition, its dynamic state (age and stage of succession), and the degree of stress imposed by natural influences and those of civilization. These constraints will now be discussed for specific ecosystems.

1. Woodlands and Dwarf-Shrub Heaths. Stands of woody plants undergo a build-up phase at the beginning of their development, when most of the plants are young. The mass of foliage must feed a relatively small mass of axes and roots, k_p is favorable, and the net primary productivity is therefore large. There is a considerable surplus of organic matter, which visibly increases the mass of the stand from year to year (Fig. 54).

The productive phase of growth gives way, with increasing age of the stand, to the mature phase, in which ΔB at first stays positive and later fluctuates about zero. This reduced rate of growth is brought about not by increased losses, but by the decline in net primary productivity as development of the community proceeds. The larger the trees grow, the smaller is the ratio of green to non-green tissues. As a result, the yield of photosynthesis suffices only for renewing the leaves and for the respiration of the enormously enlarged mass of shoot and root systems. In leafy forests the increase in wood comes to a halt when the mass of leaves falls to less than 1% of the total mass.

2. Grassland and Herbaceous Stands. In the course of the production period the phytomass grows rapidly, and at the same time parts of the shoot system and the roots die off or are eaten. In herbaceous plant communities left in their natural state, at the end of the growing season the leaves turn yellow and dry up, and parts of both the shoot and the root systems are withdrawn. In the steppes, this loss accounts for more than half of the phytomass formed during the year, and in desert plant communities consisting primarily of ephemers, it may be 60—100%. In fields that are regularly mown or grazed, the phytomass is continually removed during the growing season, so that G exceeds L. The difference between losses due to grazing and those due to shedding of foliage consists primarily in the fact that when the plants are eaten, assimilation organs are removed while still capable of full photosynthetic activity, and excessive grazing (G more than half of P_n) endangers the existence of

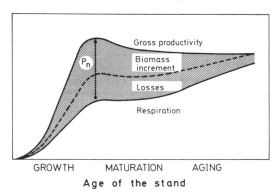

Fig. 54. Diagram of the variation, with advancing age of a uniform stand of trees, in gross productivity, net productivity (P_n, dotted area), rate of increase in biomass (dotted area above the dashed line), rate of loss (dotted area below the dashed line) and respiration. (After Kira and Shidei, 1967)

the stand. On the other hand, once the growing season is concluded, the entire phytomass above ground can die off without harm to the ecosystem.

In herbaceous stands, whether left undisturbed or under cultivation, ΔB fluctuates with steadily decreasing amplitude around zero. In dry regions with variable precipitation, ΔB fluctuates in successive years between large positive and negative values, but when averaged over many years it is also approximately zero.

3. Aquatic Ecosystems. In populations of plankton, the first effect of net primary production is an increase in the numbers of the (usually short-lived) floating algae in the euphotic zone. This supply of matter in the community serves to feed a horde of consumers; G is high, on the average $2/3$ of the net primary production. The loss L is less, but is hard to estimate quantitatively. In the case of phytoplankton, L consists of those cells that sink below the compensation depth (see p. 46), either drawn by gravity or carried by water currents. The velocity of sinking depends upon the size, shape and density of the organisms, as well as upon water movements and temperature. In cool waters at $6°$ C, most algae sink at an average rate of 3 m per day; at $20°$ C they sink twice as fast. A characteristic of aquatic ecosystems is the sequential appearance of a productive growth phase and a protective equilibrium phase (as is found in woodlands), as well as a high proportion of loss due to grazing (as is found in mown or grazed meadows). Thus in an aquatic ecosystem $G \geq L$ and the sum of the two exceeds ΔB.

Turnover of Mass and the Mobilization of Carbon

The primary producers steadily generate organic mass that first is incorporated into their own structure and later—directly or *via* the consumer chain—makes its way into the soil. Lost matter accumulates at first as litter on the ground.

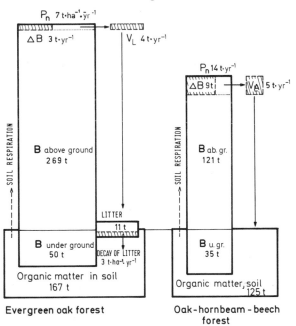

Fig. 55. Content and turnover of organic dry matter in an evergreen oak forest in southern France as compared with that in a mixed deciduous forest in Belgium. *Areas with dark outlines*: content of dry matter in $t \cdot ha^{-1}$. *Dashed lines and cross-hatched areas*: turnover rate of organic dry matter, in $t \cdot ha^{-1} \cdot yr^{-1}$. Part of the net annual production (P_n) goes to increase the phytomass of the community (ΔB), and part provides organic substances to the soil by way of the annual loss from the trees (L). (After Duvigneaud and Denaeyer-de Smet, 1970, and Rapp, 1971)

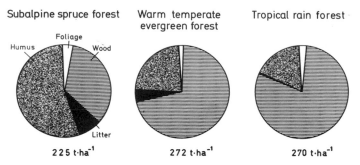

Subalpine spruce forest   Warm temperate   Tropical rain forest
                           evergreen forest

225 t·ha⁻¹   272 t·ha⁻¹   270 t·ha⁻¹

Fig. 56. Content of organic carbon in the foliage, the wood, and the ground litter and humus in woodlands of regions differing in climate. In the tropical rain forest the carbon is tied up primarily in the plant mass; the litter is rapidly converted and there is only a little organic matter in the soil. In the subalpine spruce forest, on the other hand, more organic matter is found in the litter and humus than in the plant canopy, chiefly because under the prevailing cool, damp conditions decay in the soil takes place much more slowly than in the tropics. (After Yoda and Kira, 1969)

The amount of litter in an ecosystem depends upon the total amount made available in the course of a year ($5-7$ t $\cdot$ ha^{-1} in woodland, $6-10$ t $\cdot$ ha^{-1} in grassland) and the rate of its decomposition. In tropical rain forests organic waste is decomposed within a year, but in deciduous woodlands of the temperate zone decomposition of one year's accumulation requires $2-3$ years, in coniferous woodlands $4-5$ years, and in mountain woodlands decades. In steppes decomposition occurs periodically, being rapid in spring and in summer before the dry season begins and slower later on, so that the layer of litter is deepest in winter and thin in summer. In an undisturbed ecosystem a balanced relationship between the plant mass, the amount of litter, and the quantity of organic matter in the soil becomes established; this relationship is characteristic of the species and age composition of a community and is regulated by climatic and edaphic factors (Figs. 55, 56, and 60).

By the activity of the decomposers, the litter covering the ground is gradually, through several intermediate steps, reduced to humus and finally to the inorganic level (cf. pp. 95 and 125). In the process, the CO_2 bound by the plants is once again set free.

Soil Respiration

The decomposition of organic matter and the associated CO_2 formation in the soil is promoted by an abundance of such matter, by a neutral to weakly alkaline pH of the soil, and by suitable humidity, temperature and oxygen supply. Under such conditions, the CO_2 content of the soil can rise to $0.5-1.5\%$ by volume—that is, fifty times the concentration in the atmosphere. In soils densely packed with roots, up to 10% of the CO_2 in the subterranean air comes from respiring plant roots.

Escape of CO_2 into the layer of air just above the ground is called soil respiration. Gas transport from the soil to the atmosphere—and within the soil—is the more rapid, the greater the volume of the pores in the soil and the less water they contain.

Within the soil, diffusion is the most important mechanism reducing concentration gradients; at the surface of the ground, of course, movement of the air is also involved. Soil respiration is a useful indicator of the rate of decomposition in the ground. In the temperate zone, forest floor and grass-covered soil on the average give off $0.1-1$ g $CO_2 \cdot m^{-2}$ surface area of ground per hour; poorer soils release correspondingly less.

Table 10. Net primary productivity of the earth's plant cover. (From Lieth, 1972)

Type of community	Area (10^6 km)	Net primary productivity (dry matter)		
		Range (t $\cdot$ ha^{-1} $\cdot$ yr^{-1})	Approx. mean (t $\cdot$ ha^{-1} $\cdot$ yr^{-1})	Total for area (10^9 t $\cdot$ yr^{-1})
Total continental	**149.0**		**6.7**	**100.2**
Forests	*50*		*13*	*64.5*
Tropical rain forest	17.0	10–35	20	34.0
Drought-deciduous forest	7.5	6–35	15	11.3
Winter-deciduous forest	7.0	4–25	10	7.0
Chaparral	1.5	2.5–15	8	1.2
Warm temperate mixed forest	5.0	6–25	10	5.0
Boreal forest	12.0	2–15	5	6.0
Woodland	*7.0*	*2–10*	*6*	*4.2*
Dwarf shrubs and open scrub	*26.0*		*0.9*	*2.4*
Tundra	8.0	1–4	1.4	1.1
Desert scrub	18.0	0.1–2.5	0.7	1.3
Grassland	*24.0*		*6*	*15.0*
Tropical grassland	15.0	2–20	7	10.5
Temperate grassland	9.0	1–15	5	4.5
Deserts	*24.0*		*0.01*	*–*
Dry desert	8.5	0–0.1	0.03	–
Ice desert	15.5	0–0.01	0	–
Cultivated land	*14.0*	*1–40*	*6.5*	*9.1*
Fresh water	*4.0*		*12.5*	*5.0*
Swamp and marsh	2.0	8–40	20	4.0
Lakes and streams	2.0	1–15	5	1.0
Total oceans	**361.0**		**1.6**	**55.0**
Reefs and estuaries	2.0	5–40	20	4.0
Continental shelf	26.6	2–6	3.5	9.3
Open ocean	332.0	0.02–4	1.25	41.5
Upwelling zones	0.4	4–6	5	0.2
Total for the earth	**510.0**		**3**	**155.2**

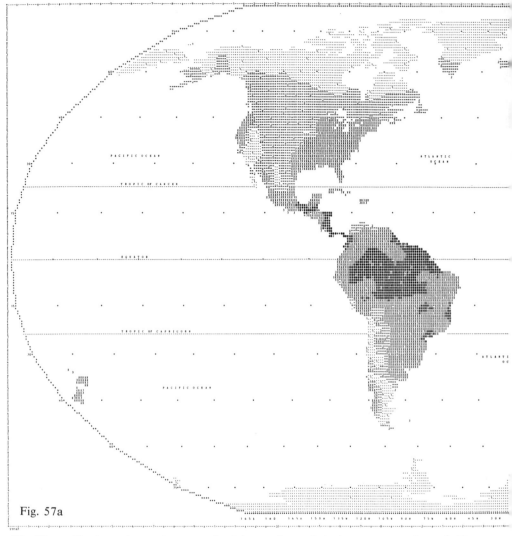

Fig. 57a

Fig. 57a and b. Annual net primary productivity on the earth (g dry matter · m⁻² · yr⁻¹). The maps were generated by computer from data obtained by H. Lieth, J. H. Lieth, Zaeringer,

Productivity of the Earth's Plant Cover

Over the whole earth, according to an extrapolation worked out by H. Lieth, the plants fix annually about $155 \cdot 10^9$ t carbon, $95 \cdot 10^9$ t (61%) of it on land and $60 \cdot 10^9$ t (39%) in bodies of water (Table 10).

High primary productivity is limited to those regions of the continents and the oceans that offer the vegetation an optimal combination of water, warmth and nutrient salts. On land this is found in the tropics and in water, in the zone between 40°

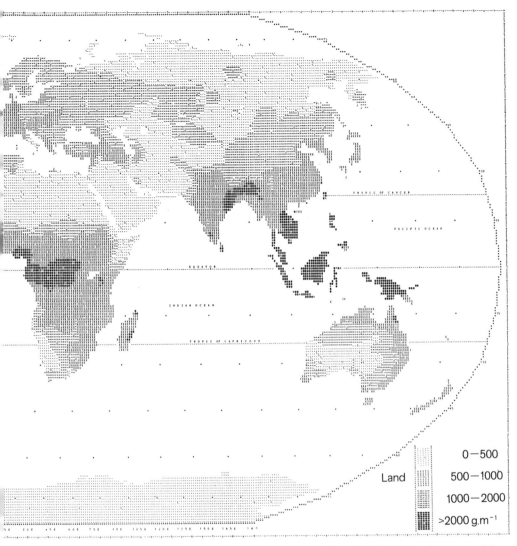

Land	0—500
	500—1000
	1000—2000
	>2000 g.m^{-1}

and Berryhill for the continents (a) and by Hsiao, van Wyk, and H. Lieth for the oceans (b). Details of the program used to produce this map are described by Lieth (1972)

and 60° north and south latitudes (Fig. 57). But the most abundant production is found in the transitional zones where land and water meet—in shallow water near the coasts and on coral reefs, in rain forests, water meadows and swamps in warm countries (Table 10 and Fig. 58). The greater part of the earth's area, both land and water, permits only moderate productivity. On the continents it is almost always the water supply that is insufficient; in the arctic and the high mountains there is in addition a shortening of the production period due to cold. In tropical seas it is lack of nutrients, and in seas near the poles lack of light, that limits productivity. On a

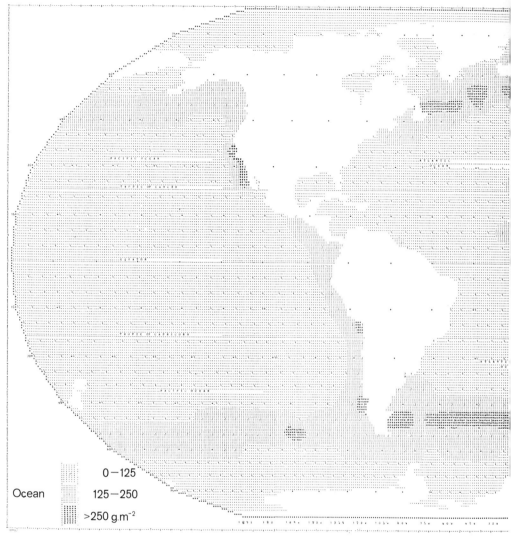

Ocean		0–125
		125–250
		>250 g.m⁻²

Legend see pp. 78 and 79

smaller scale, too, there are differences in productivity, sometimes quite pronounced, between adjacent regions; these depend on local variations in food supply, type and composition of the plant communities, and the degree of intervention by man. Intensive cultivation can achieve yields in a given area that far exceed those of the primary productivity of the plant communities that would have been found naturally in that area. On the average, though, throughout the world, the agricultural yields are far less than could be achieved, chiefly because of inadequate techniques of cultivation, extensive rather than intensive use of the land, incomplete utilization of the local production period, and failure to use the best varieties.

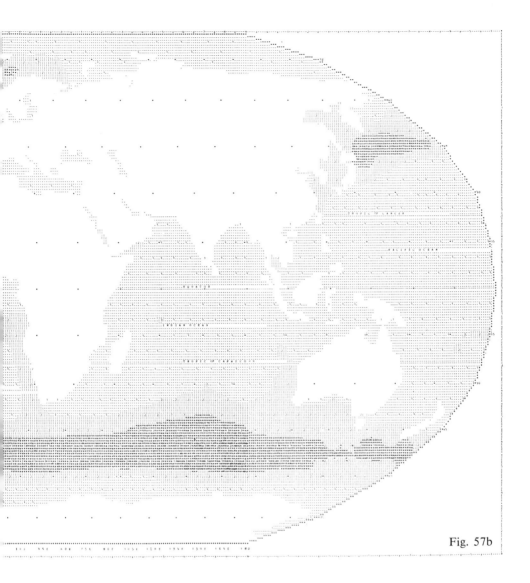

Fig. 57b

Energy Conversion by Vegetation

The Efficiency of Light Utilization in Photosynthesis

The efficiency (ε_F) of the conversion of radiant energy to chemical energy by photosynthesis is given by

$$\varepsilon_F = \frac{\text{stored chemical energy} \cdot 100}{\text{absorbed radiant energy}} \qquad (19)$$

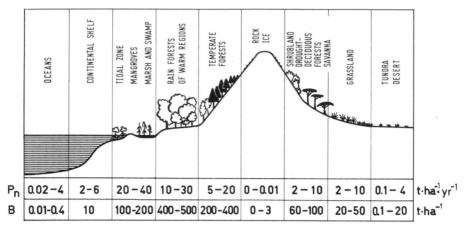

	OCEANS	CONTINENTAL SHELF	TIDAL ZONE MANGROVES MARSH AND SWAMP	RAIN FORESTS OF WARM REGIONS	TEMPERATE FORESTS	ROCK ICE	SHRUBLAND DROUGHT-DECIDUOUS FORESTS SAVANNA	GRASSLAND	TUNDRA DESERT	
P_n	0.02–4	2–6	20–40	10–30	5–20	0–0.01	2–10	2–10	0.1–4	$t \cdot ha^{-1} \cdot yr^{-1}$
B	0.01–0.4	10	100–200	400–500	200–400	0–3	60–100	20–50	0.1–20	$t \cdot ha^{-1}$

Fig. 58. Differences in primary productivity (P_n) and in phytomass (B) in different biomes over the earth. All numbers refer to t dry matter. (Based in part on E. P. Odum, 1971, with figures estimated from calculations by Whittaker, 1960; Bazilevich and Rodin, 1971; Lieth, 1972)

ε_F indicates the percent of the absorbed radiant energy fixed in the form of chemical bonds by the conversion of carbon dioxide to carbohydrates. The photosynthetic process involves binding of 9.4 kJ per gram of carbon dioxide assimilated, or 15.5 kJ per gram of carbohydrate synthesized (469.3 kJ per gram-atomic-weight of carbon). The product of this conversion factor and the gross rate of photosynthesis gives the quantity of energy bound. It goes without saying that all values must refer to the same area and time interval. The utilization of radiation by photosynthesis in single leaves under particularly favorable circumstances attains efficiencies up to 15% (with C_4 grasses, up to 24%); however, leaves usually operate at efficiencies of 5–10% or even less.

Utilization of Radiation by the Plant Cover

The efficiency coefficient (ε_P) for *primary productivity* is derived from the gross productivity of the vegetation (P_g, expressed as energy bound per year per m^2 ground area) and the amount of PhAR absorbed in the same time per unit ground area (I_{abs}).

$$\varepsilon_P = \frac{P_g \, (kJ \cdot m^{-2} \cdot yr^{-1})}{I_{abs} \, (kJ \cdot m^{-2} \cdot yr^{-1})} \cdot 100 \qquad (20)$$

If the absorbed PhAR has not been measured, it can be estimated as about 40% of the total incident short-wave radiation between 0.3 and 3 μm, for which figures are available[1]. The energy bound in the plant mass is determined from caloric measure-

---

[1] The long-term average proportion of PhAR (0.4–0.7 μm) in the total incident short-wavelength (direct + diffuse) radiation (0.3–3 μm) is *ca.* 45% (Monteith, 1965); the amount of this absorbed by the plant is about 90% (Yocum et al., 1964; cf. also Fig. 5).

Table 11. Energy content of the organic dry matter of plants. (Data from the measurements of numerous authors)

Plant material	kcal · g^{-1}	kJ · g^{-1}
Planktonic algae	4.6—4.9	19.3—20.5
Seaweed	4.4—4.5	18.4—18.9
Lichens and mosses	3.4—4.6	14.2—19.3
Herbs		
shoots	3.8—4.3	15.9—18.0
roots	3.2—4.7	13.4—19.7
seeds	4.4—5.0	18.4—21.0
Epiphytes	3.8—4.0	15.9—16.8
Deciduous trees		
leaves	3.9—4.8	16.3—20.1
wood	4.2—4.6	17.6—19.3
roots	4.0—4.7	16.8—19.7
Tropical forest trees		
leaves	3.8—4.1	15.9—17.2
wood	3.9—4.2	16.3—17.6
roots	3.9—4.2	16.3—17.6
Sclerophyllous Mediterranean woody plants		
leaves	4.8—5.2	20.1—21.8
wood	4.5—4.7	18.9—19.7
roots	4.2—4.7	17.6—19.7
Evergreen conifers		
needles	4.9—5.0	20.5—21.0
wood	4.7—4.8	19.7—20.1
Alpine and tundra dwarf shrubs		
leaves	5.0—5.6	21.0—23.5
stems	5.1—5.8	21.4—24.3
Plant constituents		
Carbohydrates	4.0—4.2	16.8—17.6
Lignin	6.3	26.4
Lipids	9.0—9.3	37.7—39.0
Terpene	11.0	46.1

ments of representative samples, and is given in terms of the caloric value. The caloric value is high if the carbon content of the organic dry matter is high, and it differs according to species and organ of the plant (Table 11) and the time of year. One may take an energy content of roughly 18.5 kJ · g^{-1} organic dry matter as typical.

Table 12. Efficiency of radiation utilization by the earth's plant cover. (Annual binding of energy from Lieth, 1972; annual irradiation from the calculations of various authors as given by Geiger, 1965)

Type of community	(1) Mean yearly energy fixation by net primary production (kcal · m^{-2} · yr^{-1})	(2) Estimated energy fixation by gross production (kcal · m^{-2} · yr^{-1})[a]	(3) Year's total radiation received at the surface of the earth (total short-wave radiation in kcal · m^{-2} · yr^{-1})	(4) Year's total photosynthetically active radiation[b]	(5) Efficiency of radiation utilization ε_p, average for the year[c]	(6) Energy utilization via net productivity[d]
Continents						
Tropical rain forests	8200	25000	1400 · 10^3	560 · 10^3	4.5%	1.5%
Winter-deciduous forests	4600	7100	1100 · 10^3	440 · 10^3	1.6%	1%
Sclerophyllous woodland	3900	6000	1500 · 10^3	600 · 10^3	1%	0.65%
Boreal coniferous forests	2400	3700	800 · 10^3	320 · 10^3	1.1%	0.75%
Tropical grassland	2800	3500	1400 · 10^3	560 · 10^3	0.6%	0.5%
Temperate grassland	2000	2300	1000 · 10^3	400 · 10^3	0.6%	0.5%
Tundra	600	900	600 · 10^3	240 · 10^3	0.6%	0.25%
Semideserts	300	350	1800 · 10^3	720 · 10^3	0.05%	0.04%
Agricultural areas	2700	3200	1100 · 10^3	440 · 10^3	0.7%	0.6%
Oceans						
Open ocean	600	670	1200 · 10^3	480 · 10^3	0.14%	0.12%
Coastal zones	1600	1800	1200 · 10^3	480 · 10^3	0.4%	0.35%

[a] Computed from the net primary productivity (column 1), using the mean productivity coefficient

[b] Fraction of the absorbed PhAR in the total incident short-wavelength radiation (column 3), ca. 40% according to Yocum et al. (1964)

[c] ε_p computed from the gross productivity (column 2) and the absorbable PhAR (column 4)

[d] Computed from the energy bound by net production (column 1) and the absorbable PhAR (column 4)

On the average, with all the varying temporal and spatial conditions affecting assimilation throughout the production period, plant communities operate with efficiencies below 2—3%, and at most, under intensive agriculture, up to 6% (in rice paddies). Relatively unproductive plant communities achieve efficiencies below 1%, and the same is true for a large part of the oceans (Table 12). In comparison with the possible values of the photosynthetic efficiency coefficient achievable, the yield under ordinary conditions is extremely meagre. The vegetation cannot make full use of all the daylight hours; it is often in those regions exposed to considerable incident radiation (near the Tropics of Cancer and Capricorn and at high altitudes) that the production period can be short. Latitudes near the Tropics, unlike those closer to or further from the equator, are characteristically high-pressure zones with little cloud cover; the sunlight provides a large energy input, but owing to the lack of rain, production is restricted and relatively little of the energy is bound. Hence a world map of the utilization of radiation by assimilation would show a narrow strip near the equator and another encompassing the temperate latitudes, representing the zones in which photosynthesis is most efficient.

The Role of Plants in the Carbon Cycle

The Distribution of Carbon on the Earth

The total carbon bound in compounds on the earth is estimated to be $26 \cdot 10^{15}$ tons. The greater part of this is inorganically bound, only about 0.05% being in organic compounds (Table 13).

Table 13. Carbon stores on the earth; units for all numbers are gigatons (10^9 t) of carbon. (From Müller, 1960; Buch, 1960; Bolin, 1970)

Inorganic carbon	
Atmosphere	700
Hydrosphere	
Inland waters	250
Ocean, surface water	500
Ocean, deep water	34,500
Lithosphere	
Coal, petroleum	7,500
Rock	25,000,000
Organically bound carbon	
Biomass (land organisms)	410
Biomass (marine organisms)	> 10
Detritus and soil (continents)	710
Detritus and organic sediments (oceans)	3,000

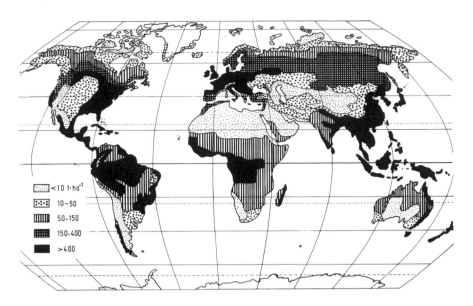

Fig. 59. Distribution of phytomass (plant mass both above and below the surface of the ground, in *t* dry matter per *ha*) on the earth. Compare this figure with the productivity map (Fig. 57) and the world map of water availability (Fig. 112). (After Bazilevich and Rodin, 1971)

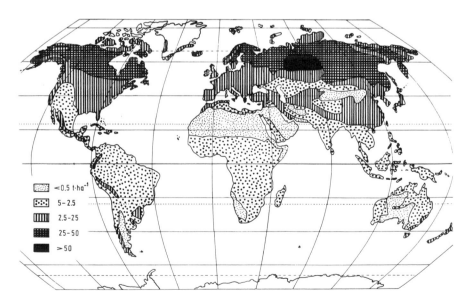

Fig. 60. Distribution of organic dry matter in ground litter and duff on the earth (cf. also Fig. 56). (After Bazilevich and Rodin, 1971)

Organically Bound Carbon

Organically bound carbon is found in the biosphere and in the upper layers of the lithosphere: 64% of this is in the form of fossil stores (peat, coal, petroleum), 32% is contained in organic detritus in the soil and the waters, and about 4% is in the structural components of the biomass. By far the greater fraction of the latter quantity is associated with land plants. Within the phytomass the forests, with their great amount of wood, predominate, storing more than three quarters of the carbon fixed by all land plants. The wooded zones of the earth are thus the regions with the largest reserves of organic carbon (Fig. 59). On the other hand, the greatest carbon stores on and within the soil are in the tundra zones and in the northern forests, where waste matter is decomposed much more slowly than in warmer regions (Fig. 60).

Inorganically Bound Carbon

The store of inorganic carbon is found largely in sediments of the earth's crust. The waters of the oceans and lakes store 0.14% of the total carbon of the earth, in the form of bicarbonate and carbonate ions or as dissolved CO_2. Most of this is in the deep waters of the ocean. The superficial layer, where the water plants live, contains only one sixtieth of the inorganic carbon of the hydrosphere. In the atmosphere there is about as much carbon as in the surface waters of the ocean; between the two there is a continual and intensive exchange of carbon. The average concentration of carbon dioxide in the atmosphere is 0.03% by volume (= 300 ppm; the mean value for 1970 was 322 ppm). At an air pressure of 1 bar, this concentration corresponds to a content of 0.6 mg CO_2 per liter of air. In industrial regions the CO_2 concentration can rise to 500 ppm. This increased concentration may be ascribed to the burning of coal, petroleum and natural gas, which returns to the atmosphere the CO_2 that it lost to photosynthesis millions of years ago.

The Cycling of Carbon

The various forms in which carbon is found differ in their mobility. These forms can be either *solid* (in organisms, the soil and the lithosphere), *dissolved* (in cell sap, the water with which organisms are saturated, the water in the ground and bodies of water), or as a *gas* (in the intercellular spaces of plants, the pores within the soil, and the atmosphere). Exchange takes place between the various sites and phases. Carbon compounds in the lithosphere become dissolved, the carbon dioxide in the water is in equilibrium with that in the air, and the carbon cycle of living organisms effects the continual circulation of carbon on several levels.

1. Carbon Circulation at the Cellular Level

Chlorophyll-containing cells draw CO_2 from the air or water and from it form carbohydrates; these may be stored, used in the synthesis of tissue, or metabolized to provide energy and to liberate the CO_2 again. The cellular carbon cycle between gross photosynthesis and respiration involves little spatial transport, and proceeds rapidly.

2. Carbon Circulation in the Plant

At the level of the organism, carbon moves from its site of uptake by the leaves (as CO_2) through the sieve-tube system to the other organs; these divert part of the carbohydrates transported to them for use in respiration and store the remainder as dry matter (the increase in which is expressed by the NAR). During the cycle, too, a relatively large fraction of the carbon taken up is soon released again as CO_2.

3. Carbon Circulation in the Food Chain

The dry matter accumulated by the plants eventually (depending on the form of plant and the organ involved) falls to the consumers—sooner in the case of herbs and the non-woody parts of plants, and more slowly with tougher structural matter. The consumers gradually decompose them, freeing the CO_2. In forests the mass of wood stands for many decades or even centuries, but in herbaceous communities the biomass becomes detritus after periods ranging from several months to some years. The biological cycling of carbon in the ecosystem often requires a considerable amount of time for the final stages of decomposition of litter by way of humus—occasionally years or decades.

4. Biogeochemical Circulation

The carbon cycle is kept in motion primarily by the processes of life. Every year, plants bind in organic compounds about 6—7% of the CO_2 available in the atmosphere or dissolved in the surface waters of the oceans and lakes. About one third of the assimilated carbon is freed by the respiration of the plant itself, and one thousandth is temporarily withdrawn from circulation by sedimentation (buried underground or underwater or deposited as peat); the rest serves as the basic food of heterotrophic organisms and thus represents the point from which the flow of energy proceeds. Respiration, fermentation and decay liberate CO_2 and replenish the earth's stores of carbon dioxide.

As a result of the intensive exchange of carbon dioxide between atmosphere and biosphere, there are *daily and yearly fluctuations in the CO_2 content of the air.* During the day the photosynthetic activity of the green plants lowers the CO_2 content of the air, and at night the liberation of CO_2 by respiratory processes raises it again. In closed plant stands the daily fluctuations in CO_2 content can amount to 25% of the mean; these can be demonstrated in the air at appreciable distances above the ground. Within the vegetation stratum, if the air is still, a characteristic concentration profile is formed. During the night CO_2 accumulates between the plants and especially near the ground; in the morning the air in the green layers of the stand becomes poorer in CO_2, and toward noon the decline in CO_2 concentration, from the open atmosphere to the top of the stand, as well as from the ground (because of soil respiration) to the foliage stratum, becomes more pronounced (Fig. 61). The flux of CO_2 toward the vegetation layer can be used to compute the photosynthetic activity of plant stands covering large areas (the "aerodynamic" method of gas-exchange analysis). Such measurements show that the plants draw the CO_2 they assimilate chiefly from the atmosphere. The contribution of soil respiration to the CO_2 supply depends upon density and height of the plant canopy, and ranges from one tenth (forests) to one fifth (agricultural crops and other low-growing communities). The concentration fluctuations during a year are less pronounced than the daily fluctuations; in the northern hemisphere the CO_2 content of the air during the growing season is lower by only 8 ppm than in the winter, and the corresponding reduction in the southern hemisphere is just 2 ppm.

Fig. 61. Daily changes in the vertical profile of CO_2 concentration in the air in a forest. By day the air around the crowns of the trees loses CO_2 as a result of photosynthesis; if there is no wind a zone of CO_2 depletion is formed (305 ppm), into which CO_2 diffuses from the atmosphere and the soil (soil respiration). At night there is a stable layering in the air, with a higher concentration of CO_2 near the ground caused by soil respiration. (After Miller and Rüsch, 1960, as presented by Bolin, 1970)

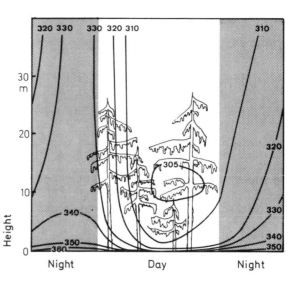

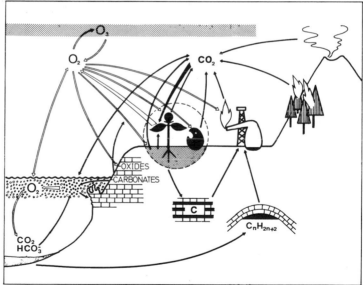

Fig. 62. Schematic diagram of the cycling of carbon and oxygen on the earth. The biological cycles (ecosystem symbols) are linked to the processes of geochemical exchange by the stores of these elements in the atmosphere and in the oceans. The burning and industrial use of coal (C), petroleum and natural gas (C_nH_{2n+2}) liberates carbon that had been organically bound millions of years ago

As compared with the biological carbon cycle, the *geochemical* turnover is negligibly small. The latter includes CO_2 release as a result of volcanic phenomena, as well as CO_2 formation in fires and industrial processes, which altogether contribute less than 10% of the quantity of CO_2 liberated annually.

5. Large-Scale Circulation

Large-scale circulation of masses of air links the land ecosystems to the CO_2 reservoir of the atmosphere. Ocean currents produce a mixing of the surface waters of the oceans and act, though very slowly, against the build-up of gradients between superficial and deep water. These two relatively independent systems, in atmosphere and ocean, are connected by the very intensive exchange of CO_2 between the air and bodies of water. Carbon dioxide dissolves readily in water, and thus the oceans of the world constitute a CO_2 buffer of enormous capacity in the global cycling of carbon (Fig. 62).

The Cycling of Oxygen

In photosynthesis and respiration the exchange of carbon dioxide and that of oxygen are coupled. The liberation of oxygen by photosynthesis maintains, in the environment of living beings, an oxygen concentration necessary for respiration. The oxygen available in air and water may well have been formed largely by autotrophic organisms, accumulating through the geological eras. The development of an oxygen-rich environment on earth was a prerequisite for the evolution of more highly organized forms of life, which have achieved their diversity of physiological performance as a result of the greater energy yield of aerobic respiration. Life on land, finally, was possible only after a shield of ozone had formed to filter out the lethal short-wave UV radiation (wavelengths below 280 nm).

The atmosphere contains about $1.2 \cdot 10^{15}$ t of oxygen; photosynthesis adds to this store $70 \cdot 10^9$ t annually. The average oxygen content of the air (21% by volume) scarcely changes, so that land plants are always abundantly supplied with oxygen. In contrast, oxygen deficiency often occurs in the soil and in bodies of water. In the soil, O_2 consumption by bacteria, fungi, animals and roots occasionally is high, but replenishment by diffusion occurs very slowly. In water, equilibrium with the oxygen pressure of the atmosphere prevails only in the uppermost layer. There the oxygen content at certain places and times can be increased as much as two- or threefold by the photosynthetic activity of phytoplankton. With increasing depth the oxygen concentration decreases because of consumption by animals, microorganisms and reducing substances in the detritus (especially H_2S). The decline is sharper, the less well mixed are the successively deeper layers of water. Thus in enclosed marine areas (the Baltic and Black Seas) and in basins where circulation is poor there can be a virtual absence of oxygen in the depths.

The cycling of oxygen on earth (Fig. 62) proceeds in a direction opposite to that of carbon. Like the latter, it consists of the two relatively independent systems in atmosphere and hydrosphere, though these again are in communication. The solubility of oxygen in water, however, is much less than that of CO_2, so that the O_2 exchange between the two reservoirs is of less significance. The real oxygen depot of the earth is the atmosphere.

The Utilization and Cycling of Nitrogen

The phytomass is not composed exclusively of photosynthates; it contains, in addition to C, O and H, an average of 2—4% nitrogen (protein contains 15—19% N). Of the bioelements, then, nitrogen takes fourth place in terms of quantity (cf. Fig. 1). Carbon metabolism and availability are closely coupled with those of nitrogen; the energy and the molecular framework for the incorporation of nitrogen come from carbon metabolism while the synthetic activity of the plant and thus the production of new tissue are controlled by the nitrogen supply. The increase in plant mass is therefore not uncommonly limited by the availability of nitrogen. Where there is a pronounced nitrogen deficiency the plants remain small, their tissues are composed of small cells and the cell walls are thickened (nitrogen-deficiency sclerosis).

The Nitrogen Metabolism of Higher Plants

Nitrogen Uptake

Green plants utilize inorganically bound nitrogen; that is, they are autotrophic with respect to nitrogen as well as to carbon. Nitrogen is taken up from the soil as nitrate or ammonium ion (for further discussion of ion absorption see p. 105). The most important source of nitrogen is NO_3^-, which is usually more abundant in the vicinity of the roots than NH_4^+. Like all ion absorptions, the uptake of nitrogen requires energy (and thus is respiration-dependent), so that in cold and poorly aerated soils plants often suffer from nitrogen deficiency.

Nitrogen Assimilation

The nitrogen taken up is incorporated into carbon compounds in amino groups, forming amino acids (Fig. 63).

The first step is the reduction of nitrate to nitrite, catalyzed by a chain of enzymes and cofactors, of which nitrate reductase plays the decisive role. Next nitrite is reduced to NH_4; this step involves nitrite reductase. The energy and the "reducing power" for assimilative nitrate reduction are provided by respiration ($NADH_2$) and—in chloroplast-containing cells—by photosynthesis ($NADPH_2$).

The actual process of assimilation is the reductive amination of α-keto acids; in higher plants the first of these is α-keto-glutaric acid, an intermediate product in the respiratory citric-acid cycle. Glutamic acid is formed, and this can transfer its NH_2 group to other α-keto acids produced in glycolysis and the citric-acid cycle (transamination). From these primary amino acids, others are derived; their carbon chains are obtained from intermediate products of carbohydrate metabolism, including the Calvin cycle and the oxidative pentose phosphate cycle. From amino acids, finally, proteins, nucleic acids and a variety of other nitrogen compounds are formed.

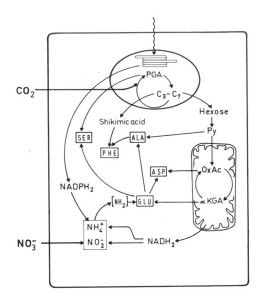

Fig. 63. Simplified diagram of nitrogen assimilation and its association with the basal metabolism of the cell. The reducing agents involved in the transfer of nitrate to the amino group are $NADH_2$ (provided *via* respiration) and $NADPH_2$ (*via* photosynthesis; i.e., photosynthetic nitrate reduction). The different amino acids arise by amination of intermediate and end products of the respiratory citric acid cycle and the dark reactions of photosynthesis. *PGA* 3-phosphoglyceric acid; *Py* pyruvate; *OxAc* oxaloacetate; *αKGA α-ketoglutaric acid; GLU* glutamic acid and related amino acids; *ASP* asparagine and related acids; *ALA* alanine and related acids; *PHE* phenylalanine and related acids; *SER* serine and related acids. Shikimic acid is derived from erythrose-4-phosphate (C_4 intermediate stage in the Calvin cycle)

Nitrogen Excretion

In the carbon metabolism of green plants the inorganic carbon source, CO_2, is taken from the air and the end product—again CO_2—is returned to the atmosphere. By contrast, higher plants take inorganic nitrogen compounds from their surroundings but break down the products of assimilation to the inorganic level (NH_3) themselves only in exceptional cases and in negligibly small quantities. Rather, the nitrogen is eliminated primarily in an organically bound form: roots release amino acids and other organic nitrogen compounds, and a greater amount of nitrogen is lost by the falling of leaves and fruits; foliage contains nitrogen in the protoplasm and also as nitrate in the cell sap, and seeds store reserve protein.

The return of the organic nitrogen compounds to the original inorganic form is accomplished by nitrogen-heterotrophic organisms (animals, many fungi and bacteria). Only through their activity is the nitrogen cycle of the higher plants completed; it is thus dependent on the nitrogen metabolism of microorganisms.

The Nitrogen Metabolism of Microorganisms

Uptake and Incorporation of Nitrogen

A large number of bacteria and fungi are capable, as are the green plants, of taking up inorganic nitrogen compounds and incorporating them in their own tissues. In this case, though, NH_4^+ is usually preferred to NO_3^-. There are microorganisms that can even make use of the exceedingly inert atmospheric nitrogen. These N_2-fixing organisms represent the most ecologically significant level of N-autotrophy. All of them are prokaryotic—cyanophytes and bacteria—some living free in the soil and others as symbionts.

Non-Symbiotic Nitrogen Fixation

In the free-living microorganisms, nitrogen fixation was first demonstrated by S. Winogradsky for the soil bacteria *Clostridium pasteurianum* and *Azotobacter chroococcum*. But there are many other species of bacteria that incorporate molecular nitrogen, chief among them the photoautotrophic bacteria and certain hydrogen-oxidizing bacteria living in water, as well as some cyanophytes of genera which form heterocysts—for example, *Nostoc, Anabaena, Calothrix* and *Mastigocladus*. These blue-green algae are self-sufficient with respect to the essential elements, being autotrophic for carbon as well as for nitrogen. Nitrogen-fixing blue-green algae are found in bodies of water and are among the first to colonize raw soils, particularly in mountains and the arctic, in thermal springs and in other extreme habitats.

Binding of atmospheric nitrogen is begun by the reductive splitting of the N_2 molecule. This strongly endergonic reaction (Table 14) is catalyzed by the nitrogenase system, a complex of two proteins, one with iron and one with molybdenum and iron

Table 14. Nitrogen-conversion reactions of microorganisms. (From Delwiche, 1970)

Reaction	kcal per mole substrate
1. Nitrogen fixation	
$\quad N_2 \rightarrow 2N$	-160
$\quad 2N + 3H_2 \rightarrow 2NH_3$	$+ 12.8$
2. Denitrification	
$\quad C_6H_{12}O_6 + 6KNO_3 \rightarrow$	
$\quad\quad 6CO_2 + 3H_2O + 6KOH + 3N_2O$	$+545$
$\quad 5C_6H_{12}O_6 + 24KNO_3 \rightarrow$	
$\quad\quad 30CO_2 + 18H_2O + 24KOH + 12N_2$	$+570$
3. Ammonification	
$\quad 2CH_2NH_2COOH + 3O_2 \rightarrow$	
$\quad\quad 4CO_2 + 2H_2O + 2NH_3$	$+176$
4. Nitrification	
$\quad 2NH_3 + 3O_2 \rightarrow 2HNO_2 + 2H_2O$	$+ 66$
$\quad 2KNO_2 + O_2 \rightarrow 2KNO_3$	$+ 17.5$

as activators. The energy and the electrons necessary for reduction are supplied by respiration. The consumption of easily accessible organic food is correspondingly high. Under laboratory conditions *Azotobacter* requires 50 g glucose per gram of nitrogen fixed, and *Clostridium* requires as much as 170 g glucose. Under natural conditions, therefore, it is to be assumed that nitrogen fixation by free-living microorganisms is limited chiefly by the supply of suitable food substrates. On raw-humus soils containing organic substances difficult to decompose, nitrogen-fixing bacteria do not thrive. In cold regions (the arctic) the rate of nitrogen fixation is also low; the nitrogenase activity of blue-green algae ceases at a few degrees above 0°C and is optimal only at about 20° C. The performance of nitrogen-fixing organisms is best in warm, permanently damp habitats; blue-green algae in rice paddies bind five times as much nitrogen as do the nitrogen-fixing soil bacteria, which on the average process 10 kg N_2 per hectare per year.

Symbiotic Nitrogen Fixation

Symbionts which fix nitrogen solve the problem of carbohydrate supply by living in the cells of autotrophic plants. In this way they manage to turn over a considerable amount of nitrogen, which is also to the advantage of the plant in which they live. Host and symbiont form an ecological unit just as self-sufficient as the blue-green algae in acquiring the necessary carbon and nitrogen compounds. The symbiotic N_2-fixing organisms bind nitrogen at a higher rate than do the free-living microorganisms; on the average one can expect fixation of 200 kg N per hectare per season, and under optimal conditions the yield is twice as great. The most important symbiotic N_2-fixing organisms are bacteria of the genus *Rhizobium*, which live with Fabaceae in nodules of the roots. There are a few species and many races of *Rhizobium*, all specialized for certain species of legumes. Another group of nitrogen-fixing symbionts comprises the actinomycetes, primarily those of the genus *Frankia*, which form root nodules in *Alnus, Myrica, Hippophae, Elaeagnus, Casuarina, Ceanothus* and certain other woody plants. All these shrubs and trees grow in nitrogen-poor soils. Blue-green algae (for example, *Nostoc* and *Anabaena*) enter symbiosis with mosses, lichens, ferns and higher plants.

Symbiosis in the Roots of Legumes. *Rhizobium* is obligatorily aerobic and lives as a saprophyte in the ground, within the range of its host plant. Plant species introduced outside their natural range (for example, soy beans, acacia, and mimosa in Europe) remain free of nodules unless the soil is first inoculated with the symbionts. Infection occurs in young plants, when the soil bacteria penetrate the root hairs and, forming an "infection thread", advance into the cortical parenchyma. There they give off a growth regulator which induces the host tissue to undergo rapid divisions, producing the nodules. At the same time the bacteria are also multiplying, fed at this stage of the infection exclusively by the host plant. Later the originally rodlike bacteria are transformed into bacteroids, which are larger and can assume branched shapes. The host cells change too, producing an enzyme related to hemoglobin ("legoglobin"), which gives them a red color. Then N_2-binding begins, catalyzed—as in the free-living organisms—by the nitrogenase system. At the time of most rapid shoot growth most of the nitrogen fixed by the bacteria is passed on immediately to the plant.

After the host plant fades the nodules age, legoglobin is broken down and the nitrogen content of the nodules falls. Finally, the bacteroids change into bacteria capable of infection. Bacteria that remain in decaying nodules are destroyed.

The Microbial Decomposition of Organic Nitrogen Compounds and the Mobilization of Nitrogen

In the detritus derived from primary and secondary producers, organic nitrogen compounds are found predominantly in the form of large molecules, chiefly protein. The molecules are split into peptides and amino acids by proteolytic enzymes (exoenzymes) secreted from the cells of fungi and bacteria; these are then taken into the cell bodies like other low-molecular-weight nitrogen compounds (for example, urea) and there undergo final decomposition. The decomposers mainly utilize the carbon chains of the molecules, the nitrogen being liberated as ammonia by oxidative desamination. The nitrogen bound in aromatic compounds is more difficult to get at, and eventually is found in the humus substances, which on the average contain 1—9% N. The mobilization of nitrogen in the soil depends upon the susceptibility to microbial attack of the fallen litter, and on its composition (particularly its nitrogen content). A characteristic measure of this composition is the ratio of carbon to nitrogen (Table 15). Substances with a very high C:N ratio are difficult for microorganisms to utilize, if no additional sources of nitrogen are available. In substrates rich in nitrogen (and thus with very low C:N ratio), on the other hand, the NH_3 losses are greater. The ratio most favorable for microbial decomposition lies between 10:1 and 20:1.

Table 15. C : N ratio. (Computed from data on the compounds making up each plant type)

Autotrophic plants	
Land plants (average)	50 : 1 to 30 : 1
Legumes	25 : 1
Planktonic algae	5 : 1 to 15 : 1
Heterotrophic organisms	
Bacteria	4 : 1
Animals	6 : 1 to 4 : 1
Large-molecule organic substances	
Plant waste[a]	
Straw	100 : 1 to 80 : 1
Leaf litter	50 : 1 to 30 : 1
Humus[a]	40 : 1 to 10 : 1
Proteins	3 : 1

[a] from Scheffer (1958)

Nitrogen Turnover in the Ecosystem

As long as man does not intervene, it is the decomposing activity of microorganisms that regulates the rate of nitrogen cycling in the ecosystem. Ecosystems left to themselves receive from other sources, by deposition or weathering, only insignificant quantities of nitrogen; the amounts carried away by water and air are slight as long as the plant cover is closed and the ground is undisturbed. The nitrogen taken from the soil by autotrophic plants in the form of NO_3^- or NH_4^+ is at first bound in the phytomass and then returned to the soil. There, nitrogen-containing organic substances accumulate until they are remineralized by ammonia-forming organisms (ammonifiers) and other decomposers. Until mineralized, the nitrogen is largely protected from leaching.

Nitrogen Mineralization

Mineralization of nitrogen proceeds slowly and in several stages. In the strict sense mineralization—that is, the transition from organic to inorganic binding—is complete once *ammonification* has occurred (Table 14). However, the ammonia produced by microbial breakdown is as a rule processed further. Many microorganisms are capable of using the various valency and oxidation levels of nitrogen (ions ranging in charge from -3 to $+5$) to obtain energy: *nitrifiers* (chemoautotrophic bacteria) oxidize NH_3 and NH_4^+ by way of nitrite to nitrate, and *denitrifiers* liberate molecular nitrogen and oxides as end products. The ammonifiers, nitrifiers and denitrifiers perform the important task of feedback control of the cycling of nitrogen in nature. By their activity, several products (NO_3^-, NH_4^+ and N_2) required by the primary and secondary producers are made available to them.

Nitrate Formation

Nitrifiers act to prevent an accumulation of NH_3 or NH_4^+ by oxidizing these two compounds. *Nitrosomonas* oxidizes NH_4^+ to nitrite, and *Nitrobacter* carries the oxidation further, to nitrate. At each oxidative step energy is obtained (Table 14) and used by these bacteria for the synthesis of organic matter. Nitrite and nitrate bacteria are always found together in the soil, since *Nitrobacter* simply takes over the end product of *Nitrosomonas*. Thus there is ordinarily no accumulation of NO_2^-. This sort of close linkage between partners in a closed food chain can be termed *parabiosis*.

Nitrifiers are sensitive to oxygen deficiency. Moreover, a high C:N ratio in the substrate used for food and a pH lower than 4.5 strongly inhibits their development. In poorly aerated and acid soils, therefore, nitrification is suppressed and NH_4^+ accumulates (Fig. 64).

The Rate of Nitrogen Mineralization

The amount of NH_4^+ and NO_3^- stored in the soil depends upon the rate of mineralization and upon other factors leading to consumption and loss of mineral nitrogen. By the rate of mineralization is meant the production of mineral nitrogen in a given

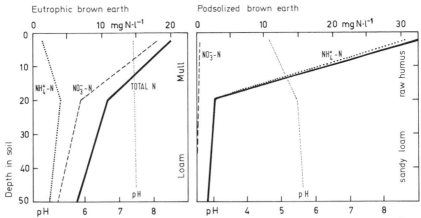

Fig. 64. Distribution of mineralized nitrogen compounds and pH in the soil of the forest floor in August (depth profile). In the acidic raw humus under a Luzulo-Fagetum, one finds NH_4^+ exclusively, whereas in the neutral mull humus under an Aceri-Fraxinetum one finds predominantly NO_3^-. (After Ellenberg, 1964)

period of time. As these inorganic nitrogen compounds are produced, higher plants remove them from the soil. But they are also used by the microorganisms for their own protein metabolism, and NH_3 and N_2 escape as a result of denitrification. The actual rate of mineralization is therefore not easy to estimate, but one can readily measure the *net rate of mineralization*—that is, the surplus mineral nitrogen remaining after the microorganisms producing it have removed what they need for themselves. The net rate of mineralization expresses the speed with which nitrogen can be returned into biological circulation within the ecosystem. Moreover, it is a measure of the nitrogen supply available to the higher plants.

The net rate of mineralization varies—even over short distances—depending on the temperature, humidity and pH in the soil as well as on the type and quantity of the nutritive substrate (cf. the differences between successive horizons in the soil profile, Fig. 64). In meadows rich in umbellifers, where nitrogen compounds are plentiful, it is very high, and in the raw humus of dwarf-shrub heaths, it is especially low. Woodlands occupy an intermediate position, coniferous forests having rates about half those of deciduous forests (Table 16). In soils with a very high C:N ratio, the net rate of mineralization is low, since the microorganisms there suffer from nitrogen deficiency and must take up inorganic nitrogen compounds. The net rate of mineralization also fluctuates considerably in time. During the year it rises and falls with changes in humidity and temperature (Fig. 65). Such fluctuations can also be brought about by increased nitrogen consumption on the part of microorganisms. For example, one can observe a decline in net mineralization in spring and summer under conditions particularly advantageous for the reproduction of microorganisms, whereas in fall and shortly after the spring thaw the rates of mineralization are high. *Ammonification* increases with rising remperature, up to 60—70° C; *nitrification* proceeds optimally between 25° and 35° C. Lower temperatures and even frost, however, do not bring nitrogen mineralization entirely to a standstill. Evidently the microflora is well able to adapt to low temperatures.

Table 16. Net mineralization rates in the upper soil strata. (From Rehder, 1970)

Type of vegetation	Average production of mineral nitrogen during the growing season (kg N · ha⁻¹)	Authors
Highly fertile meadows	300—700	Marcovič et al.
Deciduous woodland, alpine pastures, alder thickets	100—300	Ellenberg, Runge, Rehder
Coniferous forests	50—100	Zöttl
Dry grasslands, sedge moors, alpine grass heaths	20—50	Gigon, Leon, Rehder
Raw humus under alpine dwarf-shrub heath	10—20	Rehder, Ehrhardt

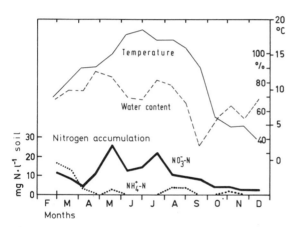

Fig. 65. Seasonal changes in nitrogen accumulation in the upper horizons of a basic brown earth under an Aceri-Fraxinetum, together with the changes in soil temperature and moisture (in %). The nitrogen content during the warm season depends chiefly upon the amount of moisture in the soil, whereas in the cold season (temperatures under 5° C) lack of warmth limits the mineralization of nitrogen, so that the water content of the soil has very little effect. (After Ellenberg, 1964)

Direct Coupling of Higher Plants to Microbial Decomposition through Mycorrhiza

In soils where mineralization is inadequate and the prevailing conditions are unfavorable to bacterial life, vascular plants can draw upon the less readily accessible sources of nitrogen. They do this with the help of the special biochemical abilities of symbiotic fungi. Almost all species of plants harbor fungal symbionts at their roots. There are two basic forms of these mycorrhiza, the ectotrophic and endotrophic types, though there are also intermediate forms. In the *endotrophic* type the mycelium penetrates the parenchyma cells of the roots, where it secretes growth regulators and other factors. In other cases, primarily woody plants, *ectomycorrhiza* are of greater significance in providing nitrogen. In this form, a dense mycelium surrounds the young (not suberized) ends of the roots, which thereupon become swollen and clublike. The hyphae also penetrate the outermost cells and the intercellular spaces of the root cortex, and take over the function of the absent root hairs. Growth of the fungus is promoted by an abundant supply of carbohydrates from the host plant.

The hyphae of the mycorrhiza, which radiate out into the soil, enlarge the absorbing surface to as much as a thousand times that of a root lacking the fungus. Through this extensive mycelial network in the soil, the fungus brings up nutrients and water and passes both on to the host plant; forest trees may obtain the greater part of their nitrogen (and phosphorus) requirement by roots with which fungi are associated. Young pines in extensive symbiosis with mycorrhiza have been shown to contain 86% more N in their needles (as well as 234% more P and 75% more K) than plants without the symbionts. The improved nutrient uptake makes possible more rapid growth in the ectomycotrophic plants and allows them to colonize unfavorable habitats such as high mountain slopes, nutrient-poor piles of industrial waste and the slag-heaps at mines.

The Role of Plants and Microorganisms in the Global Cycling of Nitrogen

The Distribution of the Earth's Nitrogen

The main storehouse of nitrogen is the atmosphere, 79% of which (by volume) is molecular nitrogen. The earth's crust, including the subsoil, contains only 0.03% nitrogen, and in the waters of the ocean the nitrogen concentration is only a few parts per million. In the topsoil, on the other hand, more nitrogen is present; it comprises 0.1–0.4% of the dry matter (cf. Table 20). The nitrogen of the soil is contained primarily in the humus, with 98% bound organically and only 2% in mineral form. Of the nitrogen on the earth that is involved in exchange processes with the biosphere, 99.4% is in the atmosphere ($3.8 \cdot 10^{18}$ t), 0.5% in the hydrosphere, 0.05% in the soil, and 0.0005% in the biomass.

The Nitrogen Cycle

Despite the large stores of nitrogen in the atmosphere, nitrogen circulates mainly between *organisms and soil* (Fig. 66). In no other cycle of a bioelement do microorganisms play such a predominant role as in that of nitrogen. Microorganisms outperform the higher plants in nitrogen metabolism. The N_2-fixing organisms are the only ones that can utilize the enormous reservoir of nitrogen in the atmosphere. They convert more molecular nitrogen into forms accessible to plants than do technological processes. According to estimates for the year 1970, about 59% of the total atmospheric nitrogen fixed on earth during the year was attributable to microorganisms, whereas industrial nitrogen preparations accounted for approximately 33%. The synthesis of nitrogen-containing organic substances by green plants and secondary producers is linked to the food chain and microbial decomposition of

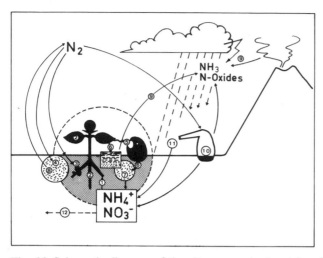

Fig. 66. Schematic diagram of the nitrogen cycle. *1* uptake of NO_3^- and NH_4^+ by microorganisms and higher plants; *2* nitrogen assimilation and the synthesis of protein; *3* fixation of atmospheric nitrogen by nodule bacteria; *4* fixation of atmospheric nitrogen by non-symbiotic nitrogen-fixing bacteria and blue-green algae; *5* provision of nitrogen to other partners in the ecosystem by way of the food chain; *6* excretion of nitrogen compounds (amino acids, urea) and other organic waste products; *7* nitrogen mineralization (ammonification, formation of nitrite and nitrate); *8* denitrification (liberation of N_2); *9* losses of NH_3 from the ecosystem and introduction of nitrogen compounds from the atmosphere by way of photo-oxidation, electrical discharges in storms and volcanic activity; *10* nitrogen binding and liberation of nitrogen compounds by industry (nitrogenous fertilizers, gaseous wastes); *11* introduction of nitrogen compounds to the ecosystem from the atmosphere through precipitation; *12* nitrogen losses from the ecosystem by leaching

detritus, and the ecosystem is linked to the atmosphere by the nitrogen-fixing organisms and the denitrifiers.

The geochemical exchange of nitrogen compounds is moderate under undisturbed conditions. Precipitation brings down to earth the ammonia, nitrate and other nitrogen oxides which are in the atmosphere as a result of electrical discharges during storms, photo-oxidation, erosion, aerosol formation, volcanic activity, and above all industrial smoke and gas discharges. The yield of these sources should not be overestimated—it can hardly exceed 10 kg N per hectare per year, and it amounts to only one seventh of the nitrogen taken into the system as a result of microbial fixation. The greatest quantities of nitrogen enter the land and water ecosystems through fertilization. This introduces to cultivated areas an average of 100 kg N per hectare per year. The loss of nitrogen by leaching is negligible where the vegetation is dense and the soil sufficiently rich in humus, since remineralized nitrogen is immediately consumed again by soil organisms and higher plants. From plowed soils, however, nitrogen is certainly washed out into lakes and streams.

The Utilization and Cycling of Mineral Elements

Plants require a large number of inorganic elements derived from minerals or mineralized by decay of organic matter (NH_4^+, NO_3^-). The mineral nutrients are taken up in the form of ions and incorporated into the plant mass or stored in the cell sap. After the combustion of the organic dry matter in the laboratory, the inorganic compounds remain as ash. In the ashes of plants, one finds all of the chemical elements occurring in the lithosphere. Some of these are essential for life; such substances include the *macronutrients* N, P, S, K, Ca, Mg and Fe, large quantities of which are required, as well as the *trace elements* or *micronutrients* Mn, Zn, Cu, Mo, B and Cl. In addition there are elements that are essential only for certain plant groups: Na for the Chenopodiaceae, Co for the Fabaceae with symbionts, Al for the ferns, and Si for the diatoms.

If metabolism is to be well regulated, production of new tissue rapid, and development unimpaired, both macronutrients and trace elements must be taken up by the plant not only in sufficient quantities but in suitable proportions. As has been known since the time of J. v. Liebig, a nutrient element available in inadequate concentration can be a yield-limiting factor. The various species of plants differ considerably in their nutrient requirements. The requirements of agricultural plants have been studied in considerable detail, but little is known about the specific needs of wild plants—even though it is precisely this sort of investigation that can provide important insights into the causes of characteristic patterns of distribution within the earth's plant cover.

The Soil as a Nutrient Source for Plants

The Mineral Nutrients in the Soil

Plant nutrients occur in the soil in both dissolved and bound form. Only a tiny fraction (less than 0.2%) of the nutrient supply is dissolved in the soil water. Most of the remainder, almost 98%, is bound in organic detritus, humus, and relatively insoluble inorganic compounds or incorporated in minerals. These constitute a nutrient reserve which becomes available very slowly as a result of weathering and mineralization of humus. The remaining 2% is absorbed on soil colloids.

Adsorptive Ion Binding and Ion Exchange in the Soil

Colloidal clay particles and humic substances, because of their surface electrical charges, attract ions and molecular dipoles and bind them reversibly. Soil colloids thus act as ion exchangers. Their exchange capacity depends upon the active surface area of the micelles; in the clay known as montmorillonite this area is 600–800 m^2 per gram, and in humic substances it can be 700 m^2 per gram or more. Both clay minerals and humic colloids have a net negative charge, so that they retain primarily cations. There are also certain positively charged sites where anions can accumulate, but there are always more cations adsorbed than anions. As a rule, the more highly charged ions are attracted more strongly—for example, Ca^{2+} more strongly than K^+—and among ions with the same charge those with little water of hydration are retained more firmly than strongly hydrated ions. By this kind of adsorptive binding, masses of ions accumulate on the surface of the much swollen micelles of clay and humus.

This coating of ions amounts to an intermediate stage between the fixed soil phase and solutions in the soil. If ions are added to or withdrawn from the soil solution, exchange takes place. Those ions that adhere more firmly are attracted more strongly by the colloid and displace other ions from its surface. The tendency for adsorption decreases in the order Al^{3+}, Ca^{2+}, Mg^{2+}, NH_4^+, K^+, and Na^+ for cations, and for anions it decreases from PO_4^{3-} through SO_4^{3-} and NO_3^- to Cl^-. Heavy metal ions can also be adsorbed, though only in trace amounts.

The adsorptive binding of nutrient ions offers a number of advantages. Nutrients freed by weathering and the decomposition of humus are captured and protected from leaching. Moreover, the concentration of the soil solution remains low and relatively constant, so that the plant roots and soil organisms are not exposed to extreme osmotic conditions (cf. pp. 120 and 139); yet when they are needed the adsorbed nutrient ions are readily available to the plants.

An equilibrium exists with respect to the soil solution, the soil colloids, and the reserves of mineral substances in the soil; it is complex and capable of adaptation. This system controls the exchange of mineral substances and ensures a continual supply of nutrient elements. The concentration of hydrogen ions in the soil solution exerts a great influence upon this ion-exchange equilibrium.

The pH of the Soil ("Soil Reaction")

Soil pH and the Availability of Nutrients

Most soils in humid regions are weakly acid to neutral, though bog soils are markedly acidic (pH about 3) and the saline and alkali soils of arid regions are basic. Acidification takes place in a number of ways: by removal of the bases through leaching, by the withdrawal from solution of exchangeable cations, by the release of organic acids by the plant roots and microorganisms, and above all by dissociation of carbonic acid, which accumulates in the soil as a product of respiration and fermentation. Depending on the parent rock and the degree of saturation of the adsorption complexes with cations, the soil is buffered to within a certain pH range. Calcareous soils are buffered primarily by the system $CaCO_3/Ca(HCO_3)_2$—that is, by the salt of a strong base and a weak acid; they are therefore weakly alkaline. The pH of the soil changes in the course of the year (especially in association with the distribution of precipitation), and there are also local differences, particularly between the different horizons in the soil (Fig. 64). Therefore to characterize a habitat

Fig. 67. Influence of the soil pH on soil formation, mobilization and availability of mineral nutrients, and the conditions for life in the soil. The width of the bands indicates the intensity of the process or the availability of the nutrients. (After Truog, 1947, from Schroeder, 1969)

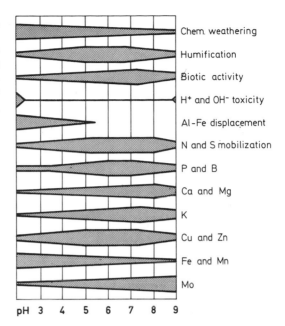

the pH must be measured over the entire year and if possible for the entire depth profile, or at least in the zone most densely penetrated by roots.

The pH of the soil has an effect upon its structure, on the processes of weathering and humification, and above all on the mobilization of nutrients and the exchange of ions. The most important of these relationships are summarized in Fig. 67. In very acid soils too much Al^{3+} is liberated, and Ca^{2+}, Mg^{2+}, K^+, PO_4^{3-} and MoO^{2-} are depleted. In more alkaline soils, on the other hand, Fe and Mn ions, PO_4^{3+}, and certain trace elements are fixed in relatively insoluble compounds, so that the plants are more poorly supplied with these nutrients.

Soil pH and the Plant

Soil pH has a direct effect on the viability of plants, in addition to its effect on the nutrient supply. Below pH 3 and above pH 9, the protoplasm of the root cells of most vascular plants is severely damaged. Moreover, the increased concentrations of Al^{3+} in very acid soils, and of borates in alkaline soils, act to poison the roots. The ranges of pH at which microorganisms can thrive are usually even narrower, and bacteria as a rule are sensitive to even moderate substrate acidity. In acid soils then, the microbial decomposition of organic substances is disturbed, decay proceeds more slowly, and there is an accumulation of NH_4^+ instead of NO_3^+ (Fig. 64). Fungi are more resistant to acid conditions and are prevalent in acid soils.

Different species display characteristic tolerance limits and requirements in their physiological behavior with respect to soil pH (Fig. 68). Some *Sphagnum* species prefer a strongly acid milieu; they are very sensitive to OH^- ions and succumb even in the neutral range. According to a classification system set up by H. Ellenberg,

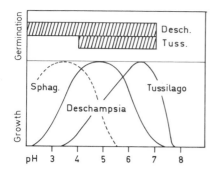

Fig. 68. Influence of the pH on the growth of a bog moss (*Sphagnum rubellum*) and on germination and growth of plants of wood hair grass (*Deschampsia flexuosa*) and coltsfoot (*Tussilago farfara*) cultivated in nutrient solution. (After Olsen as cited in Ellenberg, 1958)

they are to be regarded as strongly acidophilic, with a narrow tolerance span. The hair grass *Deschampsia flexuosa*, the presence of which is an indicator of acid soil, develops optimally at a pH between 4 and 5, but can also grow in the neutral range and will tolerate weakly alkaline soils. This species is "acidophilic-basitolerant". Similar behavior is shown by *Calluna vulgaris* and *Sarothamnus scoparius*. Converse behavior is found in coltsfoot (*Tussilago farfara*), which has an optimum in the neutral-to-alkaline range, but good tolerance down to pH 4; coltsfoot is classified as basiphilic-acidotolerant. Finally, there are extremely basiphilic organisms with a narrow tolerance span; these are not flowering plants, but various species of bacteria with an optimum tolerance far toward the alkaline end of the scale. These are damaged if the pH falls below 6.

Most vascular plants are amphitolerant, having a broad optimum in the range between weak acidity and weak alkalinity, and are able to exist between pH 3.5 and 8.5 if maintained in individual culture. In their natural range, however, some of them are restricted to a relatively narrow pH range, so that their *ecological distribution optimum* does not coincide with their *physiological development optimum* (Fig. 69). In contrast to the species which can successfully oppose competition in the range of their physiological optimum, other species are crowded out and forced into habitats where the pressure of competition is less. There they can take full advantage of the

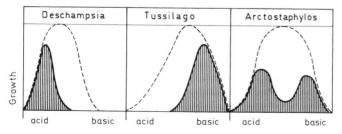

Fig. 69. Diagram of the pH-dependent growth of various plants in individual culture ("*physiological optimum curve*", dashed line) and under conditions of natural competition ("*ecological optimum curve*", solid line enclosing cross-hatched area). In the area between the dashed and solid curves the various species can thrive only if they are cultivated in a stand of that single species and are not subjected to the pressure of competition by other, better adapted species. (After Ellenberg, 1958 and Knapp, 1967)

104

breadth of their tolerance limits. In such a case it can happen that an amphitolerant species such as the bearberry (*Arctostaphylos uva-ursi*) is distributed on both acid and alkaline soils but is less common on soils of intermediate pH.

The Role of Mineral Nutrients in Plant Metabolism

The Uptake of Mineral Nutrients

The Withdrawal of Nutrient Ions from the Soil

A root takes nutrients from the soil

1. *by absorption of nutrient ions from the soil solution.* These ions are available directly, but their concentrations in the soil solution are very low: NO_3^-, SO_4^{2-}, Ca^{2+} and Mg^{2+} are present in concentrations below 1000 ppm, while there is less than 100 ppm of K^+ and less than 1 ppm of phosphate ions. The soil solution replenishes its ionic content by drawing ions from the solid phase of the soil;

2. *by exchange absorption of adsorbed nutrient ions.* By releasing H^+ and HCO_3^- as dissociation products of the CO_2 resulting from respiration, the root promotes ion exchange at the surface of the clay and humic particles and obtains in return the ions of nutrient salts;

3. *by freeing bound nutrient stores via* excreted H^+ ions and organic acids. Nutrient elements fixed in chemical compounds, chiefly heavy metals, are liberated and form chelated complexes. Metal chelates are protected from being bound again but are readily taken up by plant roots. The excretion of H^+ and acids depends upon the intensity of respiration and thus upon the availability to the roots of oxygen and carbohydrates, and upon the temperature. Furthermore, different plant species vary in their ability to make these ions available.

The Uptake of Ions by Roots

The ions of nutrient salts move from the soil solution into the parenchyma of the root cortex by diffusion or are transported by inflowing water. There they are adsorbed as a result of surface charges on the wet cell walls and at the outer boundaries of the protoplasts. This process is purely passive; it depends on the concentration gradient and the electrical potential gradient between the soil solution and the interior of the root. The outer region of the root, where equilibration of concentration with the external solution occurs by means of diffusion, is called the "*apparent free space*". Measurements on crop plants (chiefly species of grain) have shown that a considerable quantity of ions is electrostatically bound in the "apparent free space".

Ion Uptake into the Cell and the Accumulation of Ions

In penetrating living cells, ions are hindered both by their hydration and by the fact that the relevant potential gradients may oppose the motion. Uptake is therefore

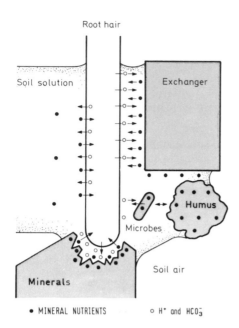

Root hair

Soil solution

Exchanger

Humus

Microbes

Soil air

Minerals

• MINERAL NUTRIENTS   ○ H⁺ and HCO₃⁻

Fig. 70. Mobilization of mineral nutrients in the soil and the uptake of mineral substances by the root. (After Finck, 1969)

accomplished only by the expenditure of energy, by systems known as "ion pumps". Energy for this *active ion transport* is obtained from respiration, so that the resulting uptake is dependent upon the factors influencing the rate of respiration. According to the "carrier" theory, the ions outside the membrane react with special binding sites on carrier molecules and are carried to the cytoplasm and released. The binding sites appear to be specific for certain ions; in any case, they discriminate between anions and cations. Pump mechanisms operate over the entire outer boundary of the cytoplasm (i.e., in the plasmalemma and in the tonoplast), and at all places within the cytoplasm where ion transport takes place between cell compartments.

The following characteristic properties of the uptake of nutrient salts can be explained in terms of both passive and active ion transport:

The Ability to Concentrate Ions. Plant cells are capable of taking up ions against a concentration gradient and accumulating them, particularly in the vacuoles, at concentrations much greater than those in the external solution. This ability is particularly important for aquatic plants, which must draw their nutrient elements from extraordinarily dilute solutions.

Selectivity. Plant cells are adapted to take up preferentially certain nutrient ions that they require. Thus cations are generally preferred to anions in the uptake process, and among the cations some are accumulated in higher concentrations than others. When necessary, electrical neutrality can be maintained by ion exchange (H^+, HCO_3^-). This selectivity is a characteristic of the physiological constitution of a plant, and its nature varies with the species.

Limits to Selectivity. Plant cells cannot entirely exclude salts that are not required or are injurious, even if the plant is damaged as a result; the biomembranes are not very permeable to ions, but the permeability is never zero. Consequently, with pronounced differences in concentration on the two sides of a membrane, some ions do leak through. Especially when the outside concentration is high—for example, in saline soils—the cells are flooded with ions (e.g., Na^+ and Cl^-) in quantities greater than are desirable.

The Translocation of Minerals in the Plant

Nutrient ions begin their movement through the plant in the "apparent free space" of the root. There they enter the cytoplasm of the parenchyma cells and move through it into the cell sap. Ion translocation from cell to cell takes place through direct plasma contact at the plasmodesmata. The vacuoles serve exclusively as storehouses, playing no role in the local salt movement. Nutrient salts that infiltrate the vacuole passively, or are actively excreted into it, remain stored there until they are actively pumped back into the cytoplasm. Thus local translocation by-passes the vacuoles and proceeeds through the *symplast* of the root—that is, through the continuous chain of living protoplasm. The symplast pathway leads through the endodermis into the core of the root. There, parenchyma cells adjoin conducting elements; the cells of the phloem take in the ions through their contact with the symplast. In the water-filled tracheae and tracheids the ions diffuse passively down the concentration gradient, and in addition they are actively excreted into the water filling the vessels.
Moving through the xylem, the nutrient salts are distributed by the transpiration stream (cf. p. 146) at higher and higher levels in the plant. At the end points of the vascular network they diffuse through the cell walls to the surface of the protoplasts of the bundle parenchyma and are actively transported into the parenchyma. Again local translocation is effected by a symplast, and again some of the salts are stored in vacuoles. In the nutrient translocation chain, the rate-limiting stage is the conduction of ions through the symplast in the roots; the transpiration stream is usually capable of carrying much greater quantities of salts.
In addition to the xylem, another important route in nutrient translocation involves the phloem. The two long-distance translocation systems are linked at many sites, particularly in the roots and in the nodes of the stems. Along with the stream of metabolites, inorganic materials are shifted to sites where the need is greatest. Translocation via the sieve tubes is involved especially in the redistribution of mineral substances already incorporated into the plant. The various substances differ in the ease with which they can be redistributed (Tables 17 and 18); nutrients bound in organic compounds, such as N, P and S, can be readily shifted, as can the alkali ions and Cl^-. More difficulty is encountered in the case of the heavy metals and the ions of alkaline earths, especially calcium. The latter thus accumulates steadily in the leaves, which mark the end of the xylem translocation route.

Utilization and Deposition of Minerals in the Plant

The inorganic bioelements may be incorporated into the plant tissues, become components or activators of enzymes, or (through effects associated with colloid

Table 17. Occurrence, uptake, distribution, incorporation and function of macronutrients. (Expanded from Finck, 1969)

Bio-element	Bound form in soil	Accessible form in soil	Taken up as	Incorporation in plant	Function in plant	Sites of accumulation	Transport-ability	Deficiency symptoms
N	organically bound, saltpeter	supplied by microbial decomposition; NH_4^+ adsorbed on clay minerals and humus; NO_3^- in solution	NO_3^-, NH_4^+, (urea)	free as NO_3^- ion (vacuoles) in organic compounds, in protein, nucleic acids sec. plant substances	essential component of protoplasm and enzymes	young shoots, leaves, buds, seeds, storage organs	good, primarily in organically bound form	stunting or dwarfism; spindly appearance; shoot: root ratio shifted toward roots; premature yellowing of old leaves
P	organically bound, phosphates of Ca, Fe, Al	as PO_4^{3-}, HPO_4^{2-}, rel. insoluble, adsorbed and in chelated complexes. Microbial release slight	HPO_4^{-2}/ HPO_4^-	free as ion, in esteric compounds, nucleotides, phosphatides, phytin	basal metabolism and synthesis (phosphory-lation)	more in reproductive organs than in vegetative (pollen granules)	good, in organically bound form	disturbance of reproductive processes (delayed flowering), stunting, bronze-violet discoloration of leaves and stems
S	organically bound, sulfur-containing minerals, sulfates of Ca, Mg and Na	SO_4^{-2} readily soluble, little adsorbed	SO_4^{-2} from soil (SO_2 from air)	free as ion, bound as SH- or SS-group and as ester, in protein, coenzymes, sec. plant substances	component of protoplasm and enzymes	leaves, seeds	good in organic form, poor as ion	similar to N-deficiency

K	feldspar, mica, clay minerals	adsorbed > dissolved	K^+	dissolved as ion (primarily in cell sap) and adsorbed	colloidal effect (promotes hydration). Synergists: NH_4^+, Na^+. Antagonist: Ca^{2+}. Enzyme activation (photosynthesis)	division zones, young tissue, bark parenchyma, sites of intense metabolism	good	disturbed water balance (tip drying), curling of leaf edges
Mg	carbonate (dolomite), silicate (augite, hornblende, olivine), sulfate, chloride	dissolved > adsorbed; deficient in acid soils	Mg^{2+}	as ion dissolved and adsorbed, bound in complexes, organically bound in chlorophyll and pectates, component of enzymes	regulation of hydration (antagonist to Ca^{2+}); basal metabolism (photosynthesis, phosphate transfer); synergists: Mn, Zn	leaves	good in part	stunted growth, interveinal chloroses of old leaves
Ca	carbonate, gypsum, phosphate, silicate (feldspar, augite)	adsorbed > dissolved; deficient in very acid soils	Ca^{2+}	as ion; as salt dissolved; crystallized and encrusted as chelate; organically bound in pectates	regulation of hydration (antagonists: K^+, Mg^{2+}); enzyme activator	leaves, tree bark	very poor	disturbance in growth by division (small cells)

Table 17 (continued)

Bio-element	Bound form in soil	Accessible form in soil	Taken up as	Incorporation in plant	Function in plant	Sites of accumulation	Transport-ability	Deficiency symptoms
Fe	sulfides, oxides, phosphates, silicates (augite, hornblende, biotite)	adsorbed >> mobilized; fixed in chalk soils	Fe^{2+}, Fe (III)-chelate	in metal-organic compounds; component of enzymes	basal metabolism (redox reactions) and nitrogen metabolism	leaves	poor	interveinal chloroses, in extreme case white coloration of young leaves (veins green)

chemistry) regulate the degree of hydration of the protoplasm. A survey of the specific ways the elements are incorporated and operate, and of the sites in the plant where they are concentrated, is given in Tables 17 and 18. Details of the biochemistry of mineral incorporation can be found in textbooks of plant physiology and microbiology.

The Incorporation of Nutrients

During the growing season, much of the total uptake and incorporation of minerals has been completed before the rapid increase in mass begins (Fig. 71). The most important nutrient elements must be made available at an early stage, and it is clear that an inadequate supply of minerals restricts the production of organic matter from the very start. The unfolding leaves of trees accumulate the chief nutritive elements—N, P, K and others—for later use. Eventually the rate of uptake of organic matter exceeds that of minerals; the ratio of organic dry matter to inorganic components begins to shift in favor of the organic matter, although the absolute quantities of mineral substances in individual leaves are not reduced. There is a decrease in absolute mineral content only if nutrients are transported out of the leaves. With increasing age, the elements Ca, S, and the other less easily moved elements Fe, Mn and B accumulate in the leaves. On the other hand, the more mobile elements N, P, and above all K are most concentrated in young leaves, their concentrations declining as the leaves mature and age. As a result, during the course of a year there is a characteristic increase of the Ca:K ratio in the leaves.

The Ash Content of Dry Matter and the Composition of Plant Ash

A survey of the ash content of various plant groups is given in Table 19; the average composition of plant ash is detailed in Table 20. There are rather large quantities (1—5% of the dry matter) of the elements N, K, Ca, and in some plants Si. Mg, P and S are found in amounts between 0.1 and 1%, and the trace-element content lies between 0.02% (Fe) and a few ppm. The proportions of the various bioelements can

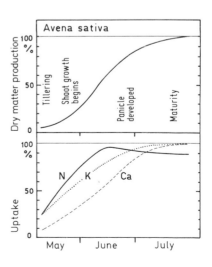

Fig. 71. Mineral uptake and production of dry matter by growing and ripening oat plants. Mineral uptake is shown as the cumulative % of the total amount taken up in the season, and the production of dry matter as the cumulative % of the total increase in dry weight for the season. N and K are taken up at an especially high rate by young, rapidly growing plants, whereas Ca-uptake essentially parallels the formation of new tissue. (After Scharrer and Mengel as cited by Mengel, 1968)

Table 18. Occurrence, uptake, distribution, incorporation and function of trace elements. (Expanded from Finck, 1969)

Bio-element	Bound form in soil	Accessible form in soil	Taken up as	Incorporation in plant	Function in the plant	Site of accumu-lation	Trans-port-ability	Symptoms of deficiency
Mn	amorphous oxide (MnO_2), carbonates, in silicates	adsorbed > > dissolved; better available in acid soils; accumulates under reducing conditions	Mn^{2+}, Mn-chelate	in metal-organic compounds and complexes; component of enzymes	basal metabolism (oxidases, photosynthesis, phosphate transfer), nitrogen metabolism; synergists: Mg, Zn	leaves	poor in part	inhibition of growth, necroses
Zn	phosphates, carbonates, sulfides, oxides, in silicates	adsorbed > > soluble; mobilization acid > basic	Zn^{2+} Zn-chelates	bound in complexes	enzyme activator, basal metabolism (dehydrogenases), protein breakdown, formation of growth regulators	roots, shoots	poor	stunted growth, discoloration of leaves, distur-bances in fructification
Cu	sulfides, sulfates, carbonates	adsorbed, mobilization acid > basic, strong fixation of humus	Cu^{2+} and Cu-chelates	bound as complexes, component of enzymes	basal metabolism (phqtosynthesis, oxidases); nitrogen metabolism; sec. metabolism	woody axes of shoots	poor	tip drying, chloroses of young leaves

Mo	molybdates, in silicates	adsorbed, mobilization basic > acid	MoO_4^{2-}	in metal-organic compounds; component of enzymes	nitrogen metabolism (reductases), phosphorus metabolism		poor	disturbance of growth and deformation of shoots
B	tourmaline, borates	adsorbed > soluble, availability acid > basic	HBO_3^{2-} $H_2BO_3^-$	bound to carbohydrates as complexes; esteric binding	carbohydrate transport and metabolism; activation of growth regulators (growth of pollen tubes)	leaves, tips of shoots	poor	disturbance of growth (meristem necroses), diminished branching of roots, phloem necroses, disturbances of fructification
Cl	salt, silicates	soluble > adsorbed	Cl^-	free as ion, mostly stored in cell sap	colloidal effect (increases hydration); enzyme activation (photosynthesis)	leaves	good	

Table 19. Average ash content of the dry matter in various groups
of plants. (From measurements by numerous authors)

Bacteria	8—10%
Fungi	7—8%
Planktonic algae without skeletal material	ca. 5%
Diatoms	up to 50%
Seaweed	10—20%
Mosses	2—4%
Ferns	6—10%
Grasses	6—10%
Herbaceous dicotyledons	6—15%
Foliage trees	
leaves	3—4%
wood	ca. 0.5%
bark	3—8%
Conifers	
needles	ca. 4%
wood	ca. 0.4%
bark	3—4%

Table 20. Average content of mineral bioelements in plants, in the soil and in sea water; all data in parts per thousand. (From compilations by Kalle, 1958; Finck, 1969; Fortescue and Marten, 1970)

Element	Land plants ($g \cdot kg^{-1}$ dry matter)		Stored in soil ($g \cdot kg^{-1}$ DM)	Marine organisms ($g \cdot kg^{-1}$ DM)	Sea water ($g \cdot l^{-1}$)
	range	mean	mean	mean	
N	10—50	20	*1*	50	*0.0003*
P	1—8	2	*0.7*	6	*0.00003*
S	0.5—8	1	*0.7*	10	*0.9*
K	5—50	10	*14*	10	*0.4*
Ca	5—50	10	*14*	5	*0.4*
Mg	1—10	2	*5*	4	*1.3*
Fe	0.05—1	0.1	*38*	0.4	*0.00005*
Mn	0.02—0.3	0.05	*0.9*	0.02	*0.000005*
Zn	0.01—0.1	0.02	*0.05*	0.2	*0.000005*
Cu	0.002—0.02	0.006	*0.02*	0.05	*0.00001*
Mo	0.0001—0.001	0.0002	*0.002*		
B	0.005—0.1	0.02	*0.01*	0.02	*0.005*
Cl	0.2—10	0.1	*0.1*	40	*19.3*

be characteristic of certain plant species and families, as well as of specific organs and ages of a plant (Fig. 72). Trees and shrubs, as a rule, contain more N than K, whereas in herbaceous plants the opposite tends to be true. The ratio Ca:K is particularly characteristic; in Caryophyllaceae, Primulaceae and Solanaceae potassium predominates, and in Fabaceae, Crassulaceae, and Brassicaceae there is more calcium. Grasses, sedges, palms and horsetails build up higher concentrations of Si than of Ca, so that this element can amount to as much as $^2/_3$ of the total ash; in diatoms, the skeletons of which are composed of silicates, Si can comprise more than 90% of the ash. Most plants contain somewhat more P than S, but the inverse ratio is also found: Brassicaceae always contain considerably more S than P. Finally, there are a number of plants that store large amounts of Na—an element normally found at the bottom of the list, just ahead of the trace elements. Chief among these plants are those growing on saline soils; these include many Chenopodiaceae as well as Brassicaceae and Apiaceae.

Within a given plant the leaves produce the most ash and the woody organs, the least (Fig. 72, Table 19). In the foliage the elements stored preferentially are Ca, Mg, S, and in the case of grasses Si as well. Flowers and fruits store mainly K, P and S, and tree trunks contain relatively large amounts of Ca.

From the content and composition of the ash one can infer—in addition to the properties of the plant species—something about the nutrient supply in the place where the plants grew. Plants on particularly nutrient-poor soil—and to an even greater extent, those on acid soils—are low in ash (1—3% of the dry matter), as are epiphytes. Conversely, plants on saline soils have a high ash content (10—25% of the dry matter), and their ash contains above-average amounts of Na, Mg, Cl and S.

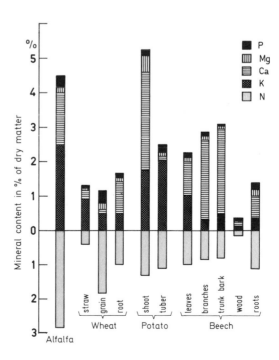

Fig. 72. Concentration and distribution of inorganic elements in different species of plants and their organs. (After Ehwald as cited by Ellenberg, 1963)

Ruderal plants, growing on nutrient-rich soils, also contain relatively large amounts of Na, as well as a good deal of nitrate stored in the cell sap. Since the plants, though able to absorb nutrient salts in the soil selectively, cannot entirely exclude any salt, the composition of the ash reflects the geochemical peculiarities of the habitat—calcium accumulation on calcareous soils, an above-average content of Fe, Mn and Al in plants from acid soils, and abnormally high concentrations of heavy metals in plants growing on serpentine soils or near ore deposits. Analysis of ash to measure habitat-dependent mineral accumulation can help in determining the presence of nutrient deficiencies and improper fertilization of crops; moreover, knowledge of their mineral content allows wild plants to be used as indicators of nutrient availability and ore deposits.

Nutrient Requirements and Excess Minerals

With respect to the quantities of inorganic elements available to plants, three basic nutritional states can be distinguished: deficiency, adequate supply, and injurious excess (Fig. 73).

When suffering from *nutrient deficiency* the plants are stunted, and in some cases they flower, bear fruits, and age prematurely. If the deficiency involves only some of the vital elements, or if the plant species requires extraordinary amounts of certain elements, *specific* deficiency symptoms can appear. These are best known for cultivated plants, but they can also be found in wild plants. The most important symptoms of specific nutrient deficiency in cultivated plants are given in Tables 17 and 18.

With an *adequate nutrient supply*, the actual amounts of the nutrients available can vary over wide ranges without noticeable effects on growth and development. Once the plant's requirements have been met, a moderate excess of certain nutrients seems to offer no further advantage for growth. But it cannot be excluded that other ecologically important properties conferring competitive advantages—such as resistance to parasites or extreme climatic situations—are in fact promoted.

In the range of *excessive concentrations*, inorganic nutrients can act as poisons, particularly if only one is present in excess. In nature such situations arise in saline and alkali soils, in ruderal habitats, and especially in soils rich in heavy metals and the rubble heaps near mines (Zn, Cr, Ni, Co, Cu). In all such places only those plants can thrive which have evolved a specific resistance to the excess metal ions;

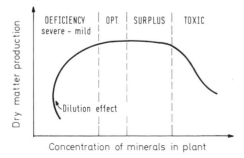

Fig. 73. Schematic representation of the relationship between inorganic nutrition (concentration of minerals in the plant) and the production of dry matter. In the range marked "dilution effect" the mineral concentration falls as dry matter is rapidly produced, though the amount of minerals is not reduced. (After Drosdoff and after Prevot and Ollagnier as cited by Smith, 1962)

such adaptations include the resistance of the calamine plants (e.g., some species of *Silene*) to Zn, the serpentine plants to Cr and Ni, *Silene vulgaris* and some grasses and mosses to Cu, *Festuca ovina* and *Agrostis tenuis* to Pb, and acidophilic grasses to Al.

The Elimination of Minerals

Most of the mineral nutrients remain in the tissues to which they have been carried. So that they will not interfere with metabolism, they are excreted into the vacuoles and stored there or precipitated (e.g., calcium oxalate). Minerals steadily accumulate in the cell walls, too, having been carried there by the transpiration stream and left as a residue when the water evaporates. These minerals are not eliminated until the corresponding part of the plant dies and is cast off, so that the abscission of leaves is a necessary and regular excretory process for perennial plants. Smaller quantities of minerals are separated out as components of secretions and excretory products, or are excreted directly (Fig. 74). In the classification scheme of A. Frey-Wyssling, *secretions* are assimilative products which are given off by the plant; examples include amino acids that seep out through the roots, and nectar. Secretions can contain minerals, which are thus removed from the inorganic metabolism of the plant. *Excretions* are defined as substances removed from the plant which are products of intermediate metabolism (growth regulators, allelopathic substances, and attractant substances such as etheric oils and amines) or end products of catabolism; these also contain small quantities of inorganic elements (primarily N, P and S).

Frey-Wyssling's scheme distinguishes as a separate category, called *recretion*, the removal from a plant of salts that have not been involved in metabolism and are eliminated while still in inorganic form. This occurs over the entire surface of the plant, and these salts are washed away by rain. The amounts eliminated in this way are not very great, but some ions (K^+, Na^+, Mg^{2+} and Mn ions) are leached out rather readily. Plants growing in saline habitats often have salt-excreting glands, and species of saxifrage excrete calcium through the hydathodes.

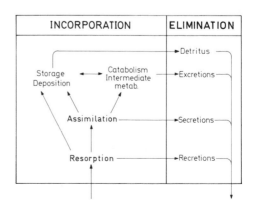

Fig. 74. Model of the turnover of inorganic matter in plants. Based in part on Frey-Wyssling, 1949

Calcicolous and Calcifugous Plants

Some species of plants can be found only on chalk soils, and others only on silicaceous and sandy soils poor in calcium. The attempt to discover the causes of this striking habitat-dependence of plant distribution was one of the first tasks undertaken by analytical ecologists. As early as the beginning of the last century F. Unger, on the basis of observations of the Tyrolean mountain flora, tried to explain the relationships involved. The problem is extraordinarily complex, and even today it cannot be regarded as solved.

Calcareous soils differ from others in the following major ways: they are usually more permeable to water and therefore dryer and warmer than silicaceous soils, but they are primarily distinguished by the fact that they contain much greater amounts of Ca^{2+} and HCO_3^-. Calcareous soils are therefore buffered toward a higher pH than other soils, and show a neutral to weakly alkaline reaction. Nitrogen is more rapidly mineralized in calcareous soils; P, Fe, Mn and the heavy metals are less accessible than in acid soils. Silicon does not appear to be involved in the question. It seems reasonable to assume that all these edaphic factors affect the inorganic metabolism and the vigor of plants, and that interspecific differences in resistance and ability to compete under a given constellation of such influences affect the composition of the plant community.

Plants with a marked preference for calcareous soils may well be damaged by iron, manganese, and above all by aluminum ions, which are liberated preferentially in acid soils. Calcifugous plants, by contrast, are capable of binding the heavy metal ions in certain complexes and thus are not harmed by a surplus of these ions. On the other hand, signs of deficiency (e.g., iron-deficiency chloroses) appear if these species are transplanted to calcareous soils, where the trace elements are present in very low concentrations.

Strict calcifuges are hypersensitive to HCO_3^- and Ca^{2+}. Peat moss and calcifugous grasses such as *Deschampsia flexuosa* produce large quantities of malate in their roots if the concentration of HCO_3^- is too great; this acts to inhibit growth and can

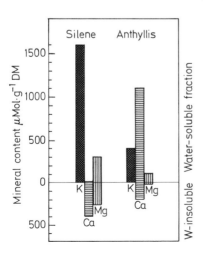

Fig. 75. Content of inorganic matter in the leaves of *Silene inflata* and *Anthyllis vulneraria* from chalky habitats. *Silene* binds the Ca taken up in water-insoluble form, as the oxalate, and contains a great deal of K. In contrast, *Anthyllis* can tolerate a high Ca level, and there is more Ca than K in the water-soluble fraction. (After Kinzel, 1969; Horak and Kinzel, 1971)

lead to root damage. Plants differ widely in their management of calcium (Fig. 75). Many Brassicaceae and some Fabaceae take up large amounts of Ca^{2+} and store it in the cell sap. Polygonaceae, Chenopodiaceae, most Caryophyllaceae and the representatives of various other plant families, on the other hand, cannot tolerate very large concentrations of dissolved calcium. Therefore they bind the Ca^{2+} in the vacuoles by way of oxalate; the more Ca^{2+} they are forced to take up, the more oxalate accumulates. Representatives of these families can maintain themselves on calcareous soils only if their acid metabolisms permit sufficient amounts of oxalate to be removed from circulation.

Plants of Saline Habitats

All saline habitats have in common a higher than normal content of readily soluble salts. The ocean, salt lakes and saline ponds are *aquatic* saline habitats; on land there are *saline soils* under both humid and arid climatic conditions. In regions of

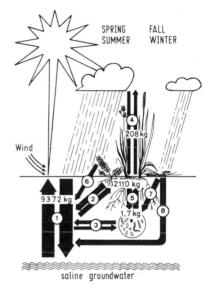

Fig. 76. Sodium-chloride circulation under humid climatic conditions in the inland habitat of a halophyte community (*Triglochin maritima, Juncus gerardi, Glaux maritima*). A saline groundwater horizon lies just under the surface of the soil; during the growing season the soil solution contains 1.9% NaCl. *1* NaCl transport in the soil, rising by capillary action during evaporation and moving down with percolating water after precipitation; *2* NaCl uptake by higher plants and release after the roots have died off; *3* NaCl uptake by bacteria in the soil and release from dying bacteria; *4* distribution and accumulation of NaCl in the parts of the plants above ground; *5* NaCl uptake and release by bacteria associated with the halophytes; *6* NaCl-leaching from living plants; *7* NaCl uptake by bacteria during the decomposition of the dead parts of plants; *8* NaCl-leaching from plant wastes and litter. Quantities are indicated in terms of kg NaCl per hectare at a depth of 10 cm in the soil. (After Steubing and Dapper, 1964)

heavy precipitation, it is possible for the soil to become salty in the spray region of the tidal zone, on dunes, and in marshes; saline soils can also be found in the vicinity of flowing waters that have been in contact with salt deposits. Moreover, the salt content of the soil can be increased if streets are salted to keep them free of ice. The salinity of the soil is greatly increased in dry areas, if the evaporation from the soil is greater in the course of a year than the amount of precipitation that infiltrates the soil (cf. Fig. 113). Especially large amounts of salt accumulate in low regions with no drainage (salt flats), as well as in places where the water table is high and irrigation is used.

Only in the open ocean does the salt content remain constant. Even in the tidal zone the salinity varies widely, and in terrestrial saline habitats the easily transported salts continually shift up and down as the water moves in the soil. During the growing season salts steadily accumulate within the plant cover as residues of evaporation; after the plant parts have died off these are washed out and returned to the soil. A survey of salt turnover and movement in a halophyte habitat is given in Fig. 76.

The Effects of High Salt Concentrations on Plants

The vigor of plants in saline habitats depends upon the concentration and the chemical composition of the soil solution. Saline soils in humid regions contain predominantly NaCl. Neutral salty soils of this sort also occur in dry regions, but more often the saline soils of steppe and desert contain sulfates and carbonates of Na, Mg and Ca, which tend to make them more alkaline. These salts affect the plants through the osmotic retention of water and the specific ionic effect on the protoplasm.

Osmotic Effect of the Salts. Salt solutions retain water, so that as the salt concentration increases water becomes less and less accessible to the plants. A 0.5% NaCl solution has an osmotic pressure of -4.2 bar, that of a 1% solution is -8.3 bar, and that of a 3% solution (the concentration of sea water) is -20 bar. Plants can draw water from the soil solution only if they can produce a negative osmotic pressure greater than that of the soil solution (for further discussion of osmotic water transport see p. 138). A first requirement for the colonization of saline soils is an ability to cope with this osmotic problem. This ability is well developed in plants exposed to large periodic changes in salt concentration—e.g. algae of the tidal zone (Fig. 77) and those living in salt pools. Some of them have developed special mechanisms for osmoregulation.

Specific Ion Effects and Salt Resistance. Once salt ions have entered the cells in high concentrations, it is the resistance of the protoplasm to the ionic effects of the salt that decides the fate of the plant. An excess of Na^+ and Cl^- causes protoplasmic structures to swell markedly and influences enzymatic activity so that metabolic disturbances can occur. Salt-sensitive protoplasts are destroyed in solutions of as little as $1-1.5\%$ NaCl, whereas salt-tolerant ones can resist up to 6% or more NaCl.

Salt resistance is a property of the protoplasm. There are extraordinarily salt-resistant organisms. The autotrophic flagellate *Dunaliella salina* lives in the concentrated saline solutions of salt pits, and the halophilic bacterium *Pseudomonas salinarum* retains functional enzymes even at concentrations of 20% NaCl. Among cultivated

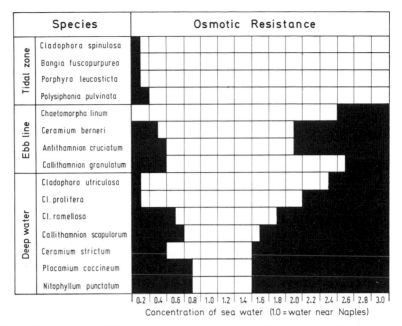

	Species	Osmotic Resistance
Tidal zone	Cladophora spinulosa	
	Bangia fuscopurpurea	
	Porphyra leucosticta	
	Polysiphonia pulvinata	
Ebb line	Chaetomorpha linum	
	Ceramium berneri	
	Antithamnion cruciatum	
	Callithamnion granulatum	
Deep water	Cladophora utriculosa	
	Cl. prolifera	
	Cl. ramellosa	
	Callithamnion scopulorum	
	Ceramium strictum	
	Plocamium coccineum	
	Nitophyllum punctatum	

0.2 0.4 0.6 0.8 1.0 1.2 1.4 1.6 1.8 2.0 2.2 2.4 2.6 2.8 3.0
Concentration of sea water (1.0 = water near Naples)

Fig. 77. Sensitivity of algae from different ocean depths to osmotic pressure of the water. White areas: surviving. Black: dead after 24 hrs. Concentration increments: "0.2" denotes 2 parts sea water + 8 parts fresh water; at "3.0" the concentration of the solutes in sea water has been tripled, by evaporation to one third of its volume. Algae from the tidal zone, which in their habitat are exposed to particularly large fluctuations of salt concentration, tolerate both hypotonic and hypertonic conditions better than the constantly wet algae of the ebb line and the sublittoral algae. (From Biebl, 1938)

plants, particularly resistant examples are barley, sugar beet, spinach, onion, rape, radish, cotton, the grapevine, mulberry trees, pomegranate, olive, date palm, acacia and various pines. Wheat, many legumes, carrots, potatoes, apple, pear, peach, apricot and lemon are very sensitive to salt. Of the trees commonly planted along city streets, the horse chestnut and linden suffer particularly from the salt spread in the winter, which contains mainly NaCl. During the growing season NaCl accumulates in the leaves and the tips of the shoots, causing necroses of the leaf margins and premature leaf fall. The limits for toxicity in these species lie at about 1.5% NaCl in the dry matter. Oaks, plane trees, and locusts are more resistant to salt and therefore better suited for curbside planting. But even those cultivated plants less sensitive to salt are by no means halophytes.

The Salt Balance of Halophytes. True halophytes (euhalophytes), according to H. Walter, are plants that live in saline habitats and accumulate relatively large amounts of salts in their organs. The salts are not harmful and in fact are beneficial as long as the quantities are not extreme (Fig. 78). Germination, in contrast, is always more successful in a salt-free milieu. It is only in the course of development that the halophyte characteristics become fully evident. As a result, seedlings are

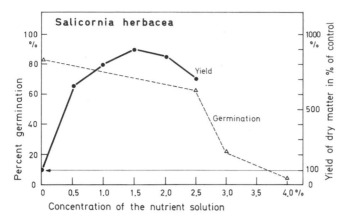

Fig. 78. Germination and yield of dry matter in the halophyte *Salicornia herbacea* as a function of salt content of the substrate. The percentage of germinating seeds was measured after 14 days. (After Feekes and Baumeister and Schmidt as cited by Kreeb, 1965)

particularly vulnerable—especially since the uppermost layers of the soil, with which they first come into contact, are the saltiest.

The special property of halophytes is a much greater than usual content of chloride and, as a rule, an excess of sodium as compared to potassium. If one grows halophytes on normal garden soil they take up Na^+ and Cl^- preferentially. Thus it is not so much a matter of their being passively flooded with these ions on saline soils, but rather that they accumulate them just as they do other actively acquired nutrient ions. This property enables them to compensate for the high osmotic potential of the soil and to draw water from the ground even if the salt concentration is high; the cell sap, of course, contains other osmotically active substances, e.g. sugar, in addition to the salts taken up (Fig. 79).

In the optimal case the salt concentration in the cell sap of halophytes just compensates the concentration in the soil solution. This is possible only if the plants can regulate the salt content of their tissues. They in fact do this by elimination of salt and by increased storage of water (succulence).

Salt Excretion. One way to regulate salt balance is the exudation of salts through salt glands and hairs. The latter are epidermal structures which actively eliminate salts and thus keep the concentration in the leaves within certain limits. They are widespread among the various mangrove plants, in species of *Tamarix, Glaux maritima*, various Plumbaginaceae, and *Triglochin*, and in halophilic grasses such as *Spartina, Distichlis* and others.

Succulence. Since the essential factor in the action of salts is not the absolute quantity but rather the concentration, a steady accumulation of salts during the growing season can be compensated if the cells also steadily draw in water, thereby becoming considerably distended. The salt *concentration* in the cell sap then remains fairly constant. Chloride ions are responsible for the development of this type of succulence. There is as yet no satisfactory explanation of the physiological processes that bring about expansion of the cells. Succulence is widespread among halophytes,

122

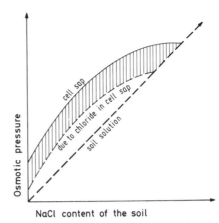

Fig. 79. Diagram of the relationship between the osmotic pressure of the soil solution (as the salt content increases) and the ability of halophytes to compensate. With increasing NaCl concentration in the soil, the osmotic pressure in the soill solution increases linearly. Halophytes overcome the correspondingly greater difficulty in withdrawing water from the soil, by taking up salt in excess if necessary and accumulating it in the cell sap (up to a certain limiting value given by the intersection of the curve showing the fraction due to chloride in the cell sap with the straight line). Added to the osmotic effect of this salt storage is that of other cell-sap components, primarily sugars (the osmotic effectiveness of non-chlorides in the cell sap is indicated by the cross-hatched region of the curve). Only when the salt concentration in the soil is very high are the halophytes no longer able to compensate for the raised osmotic pressure of the soil. (After Walter, 1960)

both in the inhabitants of damp saline habitats (*Salicornia* and other shore and salt-marsh plants in the family Chenopodiaceae) and in xerohalophytes; succulence in the latter displays additional characteristics associated with the aridity of the habitats.

Salt Elimination by the Abscission of Structures. Finally, there are very salt-resistant halophytes (e.g. *Juncus gerardi* and *J. maritimus*) that collect considerable quantities of salt in their shoots. The older leaves, full of salt, dry prematurely and fall off. Meanwhile young leaves and shoot tips capable of absorbing more salt are growing to replace them. A similar function is performed by the giant vesicular hairs of many *Atriplex* species from dry regions. These accumulate chlorides in their cell sap, then soon die and are replaced by new hairs.

Mineral Balance in the Vegetation and the Cycling of Minerals in the Ecosystem

The Mineral Balance in a Plant Community

The minerals and the carbon in vegetation are delicately balanced against one another. The uptake of minerals regulates the increase in plant mass, and carbon assimilation makes available the substance in which the minerals are incorporated. Using the production equation (16), therefore, one can determine the yearly mineral uptake by vegetation if the content and composition of the ash of the plant mass are known. A certain fraction of the total quantity of minerals absorbed (M_{abs}) by the vegetation per area of ground in the course of a year is lost during the same year in the original mineral form (recretion) or is washed out of the shoots by precipitation; this amount is called M_r. The remainder, M_i, is incorporated in the plant mass.

$$M_{abs} = M_i + M_r \tag{21}$$

The minerals fixed in the phytomass are apportioned, as in the productivity equation (17), to any increase in biomass (positive ΔB) that occurs during the year and to the annual losses of vegetable dry matter as detritus (L) and reduction by grazing (G). The quantity of minerals incorporated in each of these is computed from the fractional ash content c of the dry matter.

$$\Delta M_B = \Delta B \cdot c \tag{22}$$
$$M_L = L \cdot c \tag{23}$$
$$M_G = G \cdot c \tag{24}$$
$$M_i = \Delta M_B + M_L + M_G \tag{25}$$

The ash content c is given in kg per t dry matter, so that the annual change in the mineral content of the stand (ΔM_B) and the mineral losses M_L and M_G have the units $kg \cdot ha^{-1}$.

Within the plants, the mineral elements are unevenly distributed, accumulated, fixed and eliminated (Fig. 80). This nonuniformity should be taken into account in setting up the annual balance.

The fraction of the total mineral nutrients taken up, which is represented by M_r and M_L, can be expressed by the "*cycling coefficient*" k_M introduced by B. Ulrich:

$$k_M = \frac{M_L + M_r}{M_{abs}} \tag{26}$$

In a spruce forest in West Germany, 93% of the K taken up, over 80% of the Ca and Mg, and over 70% of the N and P were found to be returned to the soil in the same year. Values for an evergreen and a deciduous forest are given in Table 21; again potassium leads the list, followed by nitrogen and phosphorus.

In returning the greater part of its mineral uptake to the soil every year in the fallen leaves, the plant cover makes an important contribution to the circulation of minerals. These are withdrawn from deep layers of the soil by the roots, raised above the surface of the ground by the plant, and then sent back to a higher soil level. The trees in particular, because of their extensive networks of roots, raise nutrient salts that have sunk quite deep into the ground; once the litter from the trees has been broken down, the minerals are available to plants of the herbaceous stratum with shallower roots.

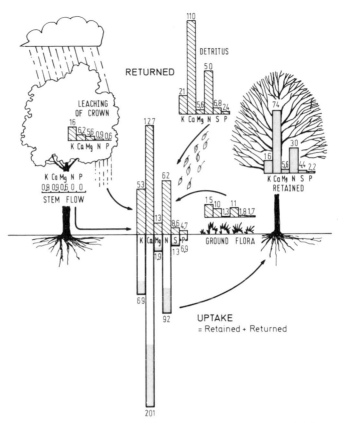

Fig. 80. The cycling of minerals in a mixed forest of oak, hornbeam and beech in Belgium. All quantities are given in units of kg · ha^{-1} · yr^{-1}. The yearly uptake of minerals by the stand was calculated by summation of the quantities retained in the phytomass (dotted areas) and those returned to the soil (cross-hatched) by way of washing and leaching of the canopy, stem flow, and the litter from trees and ground flora (cf. the carbon turnover of the same stand in Fig. 55. (After Duvigneaud and Denaeyer-de Smet, 1970)

Mineral Turnover in Land Ecosystems

The Circulation between Organisms and Soil

Under natural conditions, the cycling of minerals between the organisms in the ecosystem and the soil is, to a great extent, a closed system. The mineral nutrients taken up by the plants are returned to the soil either directly or through the food chain, packaged in organic material; there they are mineralized by the decomposers and taken up by the soil colloids. The mechanism for this *recycling* of minerals is a crucial part of the circulation; if the organic wastes (of which there are more as the productivity of the plant canopy increases) were removed, the store of nutrients represented by the incorporated minerals would be lost to the ecosystem. Recognition of this fact is the basis of well-planned fertilization; the nutrient depletion

Table 21. Mineral content and turnover in forests. Units of mineral content are kg · ha^{-1}. (Cf. Formulas 21—26 in the text)

Stand:	Deciduous oak-beech-hornbeam mixed forest with undergrowth, Belgium (Virelles), 30—75 years old Duvigneaud et al. (1969)					Evergreen oak forest, southern France (Rouquet), ca. 150 years old Rapp (1969, 1971)				
Phytomass of stand	156 t dry matter · ha^{-1}					304 t dry matter · ha^{-1}				
Annual net primary production	14.4 t dry matter · ha^{-1}					7 t dry matter · ha^{-1}				
	N	P	K	Ca	Total[a]	N	P	K	Ca	Total[a]
Mineral content of phytomass above ground	406	32	245	868	1632	763	224	626	3853	5505
Amount fixed in new growth (M_B) annually	30	2.2	16	74	127.8	13.2	2.6	8.9	42.7	68.3
Mineral content of the annual loss (M_L)	61	4.1	36	120	228.0	32.8	2.8	16.2	63.9	120.3
Mineral incorporation $M_i = M_B + M_L$	91	6.3	52.0	194.0	355.7	46.0	5.4	25.1	106.6	188.6
Leaching (M_r)	0.9	0.6	17.0	7.1	31.8	0.5	0.8	25.7	19.4	48.7
Annual absorption of minerals $M_{abs} = M_i + M_r$	91.9	6.9	69.0	201.1	387.5	46.5	5.4	50.8	126.0	237.3
Mineral cycling $k_M = \dfrac{M_L + M_r}{M_{abs}}$	0.68	0.68	0.77	0.63	0.67	0.72	0.67	0.83	0.66	0.71

[a] Total amount of mineral in the dry matter; this value is greater than the sum of N, P, K and Ca, since it includes the other elements in the ash.

brought about by harvesting of crops or removal of litter must be compensated by the provision of fertilizers in amounts of the same order of magnitude as the annual loss.

Mineralization and the production of organic matter must keep pace with one another. Where the decomposition of litter and humus proceeds too slowly, the growth rate of the plants usually slows until it is in equilibrium with the rate of mineralization. If the minerals bound in organic matter are liberated rapidly and returned in abundance, the primary producers are better supplied with nutrient ions and can build up a greater mass. The remarkable productivity of tropical rain forests is thus determined by the rapid mineral turnover there. The constant favorable temperature and the high humidity promote decomposition by microorganisms at an extraordinary rate, so that the nutrient elements in the soil are only briefly tied up in organic compounds and soon are made available to the plants again in inorganic form. Excessive liberation of the inorganic elements or a reduction in withdrawal by the vegetation (for example, after the system has been disturbed by fire or the actions of man) can disturb the mineral balance of the ecosystem. Nutrient elements in the mineralized state are not only more accessible to the plants; they are also more mobile in the soil and thus can be more easily leached out. Therefore the organically bound minerals in the biomass and in the soil are important in that they constitute relatively secure stores; all members of the ecosystem benefit.

Mineral Transport to and from the Ecosystem

The system of functional interactions represented by plants, microorganisms and soil is regularly supplied with inorganic elements from external sources, and there is a continuous loss of these elements from the ecosystem, so that the rate of microbial mineralization in the soil is not the sole determinant of mineral availability. As compared with the closed circuit of mineral turnover in the ecosystem, these geochemical exchange processes are of less importance but not entirely negligible.

The *input* of inorganic elements to the ecosystem comes from the underlying rock, the water and the air; to these are added fertilization in the case of agricultural areas. Weathering brings to the soil and the plants quantities of mineral nutrients that are hard to estimate, but no doubt significant. In raw soils it is primarily by weathering that the nutrient requirements of the plants are met. Minerals are moved into the root zone by groundwater and by water rising by capillarity. Precipitated water carries inorganic materials contained in the atmosphere as gases (SO_2, NH_3, and gases containing nitrogen oxides), dust, fog or aerosols. These atmospheric components are also captured directly by the plant cover. The supply of minerals from the air (primarily Cl, Na, Ca, S, K, Mg and N) in Europe is estimated to be $25-75$ $kg \cdot ha^{-1}$ per year.

Some inorganic elements are *lost* from the ecosystem through the atmosphere (wind erosion), but most seep down into the ground or are carried away in drainage water. The dissolved ions can re-enter a land ecosystem elsewhere, but the greater part of the leached nutrient salts eventually finds its way to inland waters and to the capacious receptacle of the ocean.

Mineral Balance of Plants in Aquatic Ecosystems

Mineral Supply and Primary Productivity

The supply of mineral nutrients to floating water plants is, next to the available radiation, the most important production-limiting factor in aquatic communities. Primary production by autotrophic planktonic organisms is possible only in the relatively narrow euphotic (adequately illuminated) layer, which becomes noticeably depleted of minerals (particularly nitrogen, phosphorus and silicon) when the phytoplankton multiplies rapidly. Even before such depletion, nitrogen and phosphorus are present only in trace amounts in fresh and sea water (Table 20). Thanks to their special ability to concentrate minerals, the planktonic algae manage initially with this limited supply. Algae can accumulate stores of phosphate so large that they suffice for synthesis and basal metabolism throughout several successive generations.

The onset of greatest productivity in waters with seasonal layering (cf. p. 19) occurs in the spring, when the higher elevation of the sun and the increase in day length permit a greater photosynthetic yield, and when the concentration of nutrients is maximal as a result of the preceding mixing of the water (Fig. 81). At this time the spring blooms of plankton occur. The growing population soon uses up the supply of nutrients; this supply is no longer adequately replenished because in the meantime the thermocline has formed and prevents the movement of mineral nutrients from the deep water into the euphotic zone.

In summer there is a reduction in the population density of phytoplankton, brought about primarily by the lowered rate of reproduction and the intensive use of phyto-

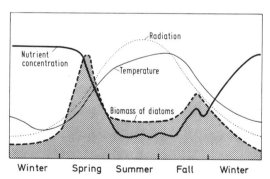

Fig. 81. Diagram of the seasonal changes in mass of a diatom population (dotted area), with the associated changes in radiation, temperature, and the concentration of mineral nutrients in the euphotic zone of the ocean. Because of the increased irradiation and warming in the spring, there is at first an adequate supply of nutrients, and production of phytoplankton mass proceeds at a high rate. The nutrients are, however, rapidly used up, which in turn soon leads to a decline in productivity. Meanwhile, the zooplankton population has greatly increased; the resulting increased consumption, together with sinking of the floating algae, in summer, reduces the biomass of the diatom population. In late fall and winter, mineral nutrients are carried up by water circulation from the deeper layers, but then it is too cold and dark for high-yield production by the phytoplankton. (After Tait, 1968)

plankton as food by the consumers, which themselves have now increased in number. Not until autumn does productivity increase, as mixing of the water is renewed; then the population density again rises somewhat before it falls back to the winter low because of lack of light.

Eutrophication and Its Consequences

Wherever mineral salts appear in the surface waters, having been washed in from the land, and wherever an upwelling of nutrient-rich deep water occurs in the oceans, eutrophic (i.e., fertilized) regions of water with greatly enhanced productivity result. In rivers and lakes, a great influx of drainage water rich in nitrates and phosphates can cause excessive fertilization which endangers the overall balance of the ecosystem. Massive development of planktonic organisms results in an excess of organic wastes and a corresponding increase in aerobic saprophytes; the latter remove so much oxygen from the deep water that this zone is rendered uninhabitable for fish and other organisms with high oxygen requirements. The biological composition of the deep-water regions of lakes is then altered fundamentally, favoring forms that can live with extremely low oxygen tension or are capable of anaerobic existence. Even in the euphotic zone the increased supply of nutrients results in direct (chiefly in the case of primary producers) and indirect (for the consumers) alterations in the qualitative and quantitative composition of the ecosystem.

In *rivers*, water plants can be damaged directly by the influx of drainage water rich in nitrates and phosphates (Table 22). If their sensitivity to increased mineral concentrations has been determined experimentally, they can be used as indicators, from their distribution, to estimate the degree of purity or pollution of the river.

Table 22. Distribution of flowering plants in a river (the Moosach in Bavaria) throughout its course, as a function of the ammonia and phosphate concentrations. (From Kohler, 1971)

	Upper reaches	Middle zone	Lower reaches
Indicator plants	*Potamogeton coloratus*[a] *Juncus subnodulosus*	*Potamogeton densus* *Hippuris vulgaris*	*Potamogeton pectinatus* *Potamogeton crispus* *Elodea canadensis* *Ranunculus fluitans* *Callitriche obtusangula*
Average content of NH_4^+ in the water ($mg \cdot l^{-1}$)	0.03	0.09	0.28
Average content of PO_4^{3-} in the water ($mg \cdot l^{-1}$)	0.04	0.07	0.45

[a] In experiments, *Potamogeton coloratus* is damaged at NH_4^+ concentrations above 10 mg · l^{-1}.

Mineral Cycling in Standing Waters

Density layering and the associated incomplete mixing of the water, prevalent during a large part of the year in lakes and in the oceans, lead to the formation of subregions with largely independent cycles of matter. Incorporation of minerals in organic substance occurs predominantly in water layers near the surface, whereas remineralization takes place in the deep water and on the bottom. The return of the inorganic products of decay to the surface water is impeded, and minerals accumulate in the depths—sometimes forming solid sediments (primarily carbonates, silicates and phosphates of calcium). These minerals are taken out of circulation for periods measured in geological terms; in such cases one speaks of a *sedimentation type* of biogeochemical cycling. Examples include the cycling of phosphorus and sulfur (Fig. 82).

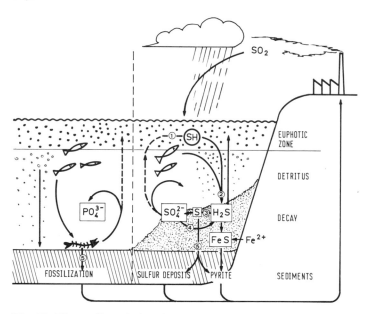

Fig. 82. The cycling of phosphorus and sulfur in an aquatic ecosystem. *Phosphorus circulation*: In the euphotic zone phosphate is incorporated into organic substances. Through the food chain, and by sinking of plankton (the rain of detritus), the phosphorus included in organic matter is transported to deeper layers of the water, which are only incompletely mixed with the surface layers. Some of the sediments are deposited on the bottom and remain there for periods measured on the geological time scale. (Based in part on Duvigneaud, 1967.) *Sulfur circulation*: The cycling of sulfur is more complicated than that of phosphorus, since many sorts of microbial transformations are involved. *1* assimilative reduction of sulfate and incorporation of sulfur in organic compounds; *2* conversion to detritus and decay of sulfur-containing organic matter; *3* aerobic microbial oxidation of H_2S by photoautotrophic and chemoautotrophic bacteria; *4* anaerobic catabolic reduction of sulfates by desulfurizers; *5* deposition of elementary sulfur and of inorganic sulfur compounds. The utilization of sulfur-containing sediments and fossil fuels (coal and petroleum) sends sulfur into the air in the form of SO_2; from the air it returns to the soil and to bodies of water, through precipitation. (Based in part on E. P. Odum, 1971)

The Phosphorus Cycle

Phosphate taken up by primary producers moves, *via* the food chain or with sinking planktonic algae, into the poorly illuminated deep water and settles on the bottom, where both algae and animal bodies decay. The phosphate incorporated in the skeletons of the animals breaks down very slowly and forms sediments. The remaining phosphate liberated is in part precipitated by cations such as calcium and iron, or adsorbed. Some, however, stays in the deep water in dissolved form, and is carried to the surface by upwellings and other currents.

The Sulfur Cycle

The cycling of sulfur in bodies of water is also of the sedimentation type, but the diversity of valency and oxidation levels of sulfur (as in the case of nitrogen) makes possible a complex involvement of various decomposers. As a result, the complete mineralization of organic wastes occurs even under unfavorable conditions, and sulfur is rapidly returned to a form that plants can take up. Hydrogen sulfide is produced by the breakdown of proteins and serves as hydrogen donor for photosynthesis by red sulfur-bacteria (Thiorhodacea) and as a substrate for chemosynthesis by sulfur-oxidizing bacteria. The chemoautotrophic bacteria (for example, *Beggiatoa*) oxidize H_2S via elementary sulfur to SO_4^{2-}, which goes into solution, distributes itself in the water, and is taken up by green plants. In addition, sulfate serves as a source of oxygen for respiration by the desulfurizers. These bacteria live in the oxygen-poor, decaying mud on lake bottoms, where they break down organic detritus; thus they act as mineralizers in an environment in which only a few organisms can exist. Under the reducing conditions in the mud, H_2S accumulates and either escapes as such or is bound to iron and deposited as pyrite. In the latter case sulfur is removed from circulation for long periods. Moreover, sulfur is deposited both in elemental form and in non-mineralized organic residues. Sedimented sulfur reenters the cycle when ores, coal and petroleum are brought to the surface. Through smelting and the burning of fossil fuels, sulfur enters the atmosphere rather than the soil or the hydrosphere. This interference by man in the cycling of sulfur is, in the long run, detrimental; the natural cycle is broken, and sulfur is no longer returned to the part of the environment from which it was taken.

The Toxic Effect of Atmospheric Pollutants

Sulfur dioxide, gases containing nitrogen oxides, the haloid acids and other air pollutants invade plants by gas exchange as well as through fog, rain and dust deposits on their surfaces.

Flowering plants exposed to large airborne concentrations of toxic materials (e.g., more than 1 ppm SO_2 or 0.1 ppm HF) suffer necrosis of some parts of the leaves, even after a short time. Even more common are chronic injuries resulting from prolonged exposure to lower concentrations (0.05–0.2 ppm SO_2). These injuries include inhibition of photosynthesis, and more frequently, paralysis of the guard cells, so that they no longer reliably control transpiration (Fig. 83). In trees, the

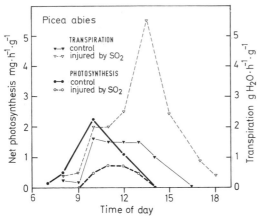

Fig. 83. Daily changes in net photosynthesis (thick line) and in transpiration, in an SO_2-damaged spruce (open symbols, dashed lines) and in an undamaged control. In this case, SO_2 poisoning brings about a paralysis of the stomata in the open position (increased transpiration) and simultaneous injury to the assimilation parenchyma (reduced photosynthesis); but plants may also react in other ways. (After Koch as cited by Th. Keller, 1971b)

mechanism for raising water from the ground breaks down, and with it the overall water balance; the leaves are cast off, and the shoots at the top dry up. With moderate exposure to toxic materials, the trees do not die, but there is a noticeable deterioration in synthesis of new substances and in growth. The growth of young shoots is diminished in trees subjected to airborne toxins, and the annual growth rings in the wood become distinctly thinner.

The various plant species are differentially sensitive to noxious gases. Among the herbaceous crop plants clover suffers the worst damage from SO_2. Some varieties of tulip and gladiolus react to HF with such sensitivity that they can be used as indicators of the gas. The woody plants exhibit wide differences in resistance (Table 23). Studies of resistance to air pollution are of practical significance in the selection of plants for industrial areas and regions of high population density, as long as account is taken of the nature of the exposure. The danger is greater, for example, in the case of evergreen trees; their foliage is exposed to the noxious gases throughout the year, particularly during the winter when buildings are heated and the amount of expelled SO_2 is increased.

Lichens are extraordinarily sensitive to SO_2, particularly foliose species (Table 24). As little as one per cent of the SO_2 concentration harmful to higher plants produces in lichens respiratory disturbances, the breakdown of chlorophyll, and inhibition of growth. From the damage observed in samples of bark overgrown with lichen (lichen explants) and from the composition of the natural lichen growth on trees and stones, it is possible to make inferences about the effects upon a habitat of long-term exposure to SO_2. In areas that receive maximal SO_2 exposure, lichens do not survive; the area becomes a lichen desert. With increasing distance from the SO_2-emitters, resistant crustaceous lichens begin to appear; these are followed by lichen communities comprising many species. Only in the undisturbed regions, however, is there again a luxuriant growth of lichens on tree trunks and rock surfaces.

Table 23. Sensitivity of woody plants to noxious gases at concentrations of 0.5–2 ppm (SO$_2$) and 0.3–0.5 ppm (HF); the gradation of the responses is based on externally visible damage. (After Ranft and Dässler, 1970, and Dässler *et al.*, 1972)

Sensitivity	to SO$_2$	to HF
Very sensitive	*Pinus sylvestris* *Larix decidua* *Picea abies* *Salix purpurea*	*Juglans regia* *Vitis vinifera* *Berberis vulgaris* *Pinus sylvestris* *Picea abies* *Larix decidua*
Sensitive	*Salix fragilis* *Salix pentandra* *Berberis vulgaris* *Rubus idaeus* *Tilia cordata* *Vitis vinifera* *Pinus nigra*	*Tilia cordata* *Rubus idaeus* *Carpinus betulus* *Pinus nigra*
Very resistant	*Juniperus sabina* *Thuja orientalis* *Buxus sempervirens* *Ligustrum vulgare* *Quercus petraea* *Platanus acerifolia*	*Chamaecyparis pisifera* *Acer campestre* *Acer platanoides* *Evonymus europaea* *Quercus robur* *Sambucus racemosa*

Additional data on sensitivity to noxious gases in various woody plants and herbaceous species are found in Garber (1967), Krüssmann (1970), and Treshow (1970).

Table 24. Sensitivity of lichens and mosses to SO$_2$ (0.6–3 ppm); the qualitative measure of the responses is based on externally visible damage (bleaching). (From Dässler and Ranft, 1969)

Sensitivity	Lichens	Mosses
Very sensitive	*Parmelia furfuracea* *Ramalina farinacea*	*Sphagnum* species *Polytrichum commune* *Polytrichum juniperinum*
Sensitive	*Cetraria islandica* *Parmelia physodes* *Cladonia alcicornis* *Cladonia arbuscula* *Cladonia rangiferina*	*Plagiothecium undulatum* *Atrichum undulatum* *Dicranum scoparium* *Hypnum cupressiforme* *Hylocomium splendens*
Resistant	*Rhizocarpon geographicum* *Lecanora varia* *Umbilicaria* species	*Pohlia nutans* *Dicranella heteromalla*

Water Relations

Life evolved in water, and water remains the essential medium in which biochemical processes take place. Protoplasm displays the signs of life only when saturated with water—if it dries out, it does not necessarily die, but it must at least enter a state in which vital processes are suspended.

Plants are composed mainly of water. On the average, protoplasm contains 85—90% water, and even the lipid-rich cell organelles such as chloroplasts and mitochondria are 50% water. The water content of fruits is particularly high (85—95% of the fresh weight), as is that of soft leaves (80—90%) and roots (70—95%). Freshly cut wood contains about 50% water. The parts of plants having the least water are ripe seeds (usually 10—15%); some seeds with large stores of fat contain only 5—7% water.

Poikilohydric and Homoiohydric Plants

In land plants, the assimilative parts of which are in contact with air and continually lose water by evaporation, the establishment of suitable *water relations* is the first requirement. Depending on whether or not they can compensate for short-term fluctuations in water supply and rate of evaporation, terrestrial plants may (according to H. Walter) be distinguished as poikilohydric or homoiohydric. Like lifeless protein gels, poikilohydric plants match their water content with the humidity of their surroundings (they are good imbibants; see Fig. 84). Bacteria, blue-green algae,

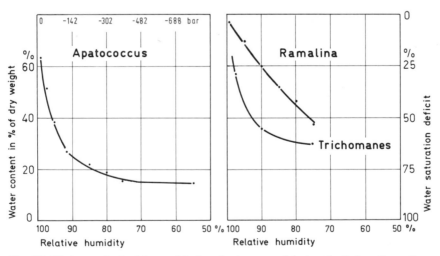

Fig. 84. Water content of the aerial alga *Apatococcus lobatus,* the lichen *Ramalina maciformis,* and the hymenophyllacean *Trichomanes radicans* in equilibrium with air at different humidities. (After Härtel, 1940; Bertsch, 1966; Lange, 1969)

lower green algae in the order Protococcales, fungi and lichens all have small cells that lack a central vacuole; when they dry out, these cells shrink very uniformly, without disturbance of the protoplasmic fine structure—the plant remains viable. As the water content decreases, the vital processes—e.g. photosynthesis and respiration (cf. p. 53)—are gradually suppressed. When sufficient water has been imbibed, such plants resume normal metabolic activity and growth. Poikilohydric plants thus have an advantage wherever there are frequently alternating periods of dryness and moisture. The humidity threshold for activity is species-specific and determines the preferred range of distribution of the various species. Soil bacteria and fungi, in general, are active only when the relative humidity (RH) is above 96%, but many molds can germinate and grow at RH between 75% and 85%; there are also fungi (for example, species of *Xeromyces*) that begin to grow at RH as low as 60%.

Poikilohydric forms are found not only among thallophytes, but also among mosses of dry habitats, certain vascular cryptogams, and a very few angiosperms. Pollen grains, and embryos in seeds, are also poikilohydric.

Homoiohydric plants are descended from green algae with vacuolated cells; a large central vacuole is the common characteristic of all homoiohydric plants. Because the water content is stabilized within limits by the water stored in the vacuole, the protoplasm is less affected by fluctuating external conditions. However, the presence of a large vacuole also results in the loss of the cell's ability to tolerate exsiccation. Thus one finds the predecessors of homoiohydric land plants pressed close to wet

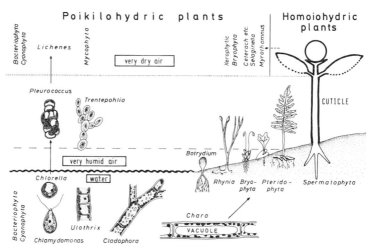

Fig. 85. Adaptation of the water economy of plants to terrestrial life. From left to right: Transition from aquatic lower algae with non-vacuolated cells to primarily poikilohydric aerial algae; development of vacuole in aquatic green algae and Characeae; transition from vacuolated thallophytes to homoiohydric vascular plants (hygrophytic mosses are still restricted to habitats with high air humidity, and in dry habitats become secondarily poikilohydric; there are also secondarily poikilohydric forms among the pteridophytes and angiosperms, but not among gymnosperms). Most vascular plants, because they are equipped with a cuticle that limits transpiration and because their cells are considerably vacuolated, are homoiohydric. (From Walter, 1967)

soil or living in permanently moist habitats (Fig. 85). Only with the evolution of a protective cuticle to slow down evaporation, and of stomata to regulate transpiration, were plants able to control their water economy adequately; with these adaptations, and an extensive system of roots, the protoplasm can be maintained in steady activity despite sudden changes in humidity. Thus the plants were sufficiently productive to form a closed cover over broad areas—and ultimately the enormous phytomass now clothing the continents.

Water Relations of the Plant Cell

The Water in the Cell

The water in plant cells occurs in several forms: It is a chemically bound constituent of protoplasm; as water of hydration, it is associated with ions, dissolved organic substances and macromolecules, filling the gaps between fine structures of the protoplasm and of the cell wall; it is stored as a reserve supply in cell compartments and vacuoles; finally, as interstitial water it serves as a transport medium in the spaces between cells and in the conducting elements of the vessel and sieve-tube systems.

Water of Hydration

In accordance with their dipole nature, water molecules aggregate and accumulate at charged surfaces in flexible arrays (known as structured water). Strongly charged ions of about the same size as water molecules bind the latter more firmly, the greater their charge and the smaller the radius of the ion. The sodium ion, with the same charge but half the surface area of the potassium ion, has twice the "density" of charge and accordingly almost four times as thick an envelope of hydration. Water molecules are bound tightly at the surfaces of the ions by electrostatic forces. A similar situation occurs at the surfaces of protein molecules and polysaccharides. Water molecules become associated with polar groups (hydroxyl, carboxyl and amino groups) and form several layers of structured water; the water molecules are more readily displaceable the further they are from the polar group.

Water of hydration accounts for only 5—10% of the total cell water, but this amount is absolutely essential to life. Only a slight decrease in the content of water of hydration results in severe alterations in protoplasmic structure and in the death of the cell.

Most of the water in the protoplasm fills the spaces in the cytoplasm and cell wall, but it is not entirely free to move; its mobility is limited by capillary forces, and it is retained by dissolved substances. The cell walls of plants hold water with a "pull" of 15—150 bar, depending on the density with which the fibers are packed. The surface forces holding water to structural elements in a matrix (cell wall, plasma colloids) can be expressed in terms of the "matric" pressure τ.

The most easily translocated water is that in those cell compartments specialized as reservoirs for solutions. But even this water is not completely mobile; in addition to hydration and matric effects, dissolved substances such as sugars, glycosides, organic acids and ions affect the net diffusion of water, through semipermeable membranes, from one solution to another. This effect is described in terms of the *osmotic pressure* of each solution.

If two solutions (for example, cell sap and pure water) having different concentrations of various solutes are separated by a membrane permeable only to water, there is a tendency for net flow of water to occur into the solution having the greater solute concentration. This is not paradoxical, and can be understood in the sense that the water diffuses *down* its *own* concentration gradient. If now the more concentrated fluid is confined (as by the cell wall enclosing the protoplast), the inflow of water causes the pressure in the cell to rise; it rises until the pressure difference just counterbalances the tendency of water to flow in. Although this equilibrium pressure difference depends upon the concentrations of the *two* solutions, one speaks of the osmotic pressure of a single solution, implying the pressure that will just counterbalance the water influx which would occur were it in communication with pure water, through a membrane permeable to water.

The osmotic pressure of a solution increases with absolute temperature and with the concentration n of dissolved molecules (the molarity in terms of molecules and ions that do not further dissociate). Van't Hoff pointed out in 1855 that, at least for dilute solutions, the osmotic pressure π is given approximately by

$$\pi = nRT \tag{27}$$

Here R is the universal gas constant (8.315×10^7 ergs $\cdot$ mole^{-1} deg^{-1}), so that a 1 M solution at standard conditions has an osmotic pressure of about 22 bar.

Solutions of dissociating substances contain more independent particles than their molarity indicates, so that for electrolytes the equation includes another factor taking into account the degree of dissociation. Conversely, macromolecular substances can be prominent constituents, in terms of weight, without raising the osmotic pressure appreciably. Through the polymerization of small molecules (with marked osmotic effect) to macromolecules—for example, the conversion of sugar to starches—and the reversal of this procedure by hydrolysis, the cell can rapidly alter its osmotic pressure; as a result, the net influx of water can be regulated.

The Water Potential of Plant Cells

The binding of water to macromolecular structures and dissolved substances results in a decreased availability for chemical reactions and for use as a solvent. It is the availability of water, rather than its total quantity, that influences the biochemical activity of protoplasm.

The thermodynamic state of the water in a cell can be compared with that of pure water, and the difference expressed in terms of potential energy. The water potential Ψ, as defined by R. O. Slatyer and S. A. Taylor, is equal to the difference in free energy per unit volume between matrically bound, or pressurized, or osmotically constrained water and that of pure water. The dimensions of water potential are energy per unit mass or per unit volume (erg $\cdot$ g^{-1} or erg $\cdot$ cm^{-3}). Because a dyne-cm is an erg, force per unit area and energy per unit volume have the same dimensions, the conversion factor being approximately 10^6 erg $\cdot$ cm^{-3} = 1 bar.

The terminology is somewhat awkward since Ψ is ordinarily a negative quantity (increased solute concentrations and decreased hydrostatic pressure lower the free energy of water), whereas similar formulations in use until recently have involved positive terms—diffusion pressure deficit (DPD) in America, and suction pressure S in Europe. The term *Hydratur*, common in the German literature, also refers to the activity of water and is closely related to water potential. P. E. Weatherley has suggested the use of "depression of water potential" $\Delta\Psi$, which is then a positive quantity. The preceding variables are related as follows:

$$S = DPD = \Delta\Psi = \pi\text{-}P \qquad\qquad (28)$$

where π is the osmotic pressure of an enclosed solution and P is the pressure exerted by the enclosing wall.

An alternative terminology is based on the consistent use of potential terms: the water potential Ψ (negative), the osmotic potential Ψ_π (negative), the matric potential Ψ_τ (negative), and the pressure potential Ψ_P (positive). This terminology will be adopted in the following sections.

Water Potential and the Cellular Translocation of Water

The rules governing water translocation in plant cells were worked out, in their essentials, almost a hundred years ago by W. Pfeffer and D. A. de Vries.

Table 25. Values of relative humidity (in the air above a solution) and the osmotic pressure (in bar) of the solution, when the two phases are enclosed and allowed to equilibrate at 20° C. (Recomputed from Walter, 1931)

bar	% RH	bar	% RH
0	100	97.9	93.0
6.7	99.5	112	92.0
13.5	99.0	126	91.0
20.3	98.5	141	90.0
27.2	98.0	301	80.0
34.1	97.5	481	70.0
41.0	97.0	687	60.0
55.0	96.0	933	50.0
69.1	95.0	∞	0
83.2	94.0		

The water balance of cells is of course intimately related to the availability of water in their environment. For a cell in air, as Table 25 indicates, there is an equilibrium between the humidity and the water potential of the cell; if the relative humidity of the air is less than the given equilibrium value, water will be lost and the water potential will become more negative. Within a cell, too, water shifts from one cell compartment to another until an equilibrium is reached, at which point the water potentials of the different compartments are equal. The processes involved in establishing these equilibria can be most simply described on the basis of the familiar model of a single cell in hypotonic and hypertonic media, as commonly found in textbooks on plant physiology (Fig. 86).

The Single Cell in a Hypotonic Medium. If a cell with a non-rigid wall is placed in a hypotonic medium, water enters the cell; it flows through the protoplasm into the cell sap, and as a result the volume of the vacuole increases considerably. The expansion of the vacuole presses the plasma against the cell wall, which stretches elastically and exerts a counterpressure (the turgor pressure P). The tendency for water to enter the cell is determined by the difference in water potential between the cell contents and the external medium, the water potential of the cell being given by

$$\Psi_{cell} = \Psi_{\pi} + \Psi_{P} \qquad (29)$$

As water enters the cell, Ψ_{π} rises (becomes less negative, as the cell solution becomes diluted) and Ψ_{P}, due to the restoring force of the cell wall, also increases. Ψ_{cell} thus becomes increasingly less negative, until it equals the water potential of the hypotonic medium.

If the cell is in pure water, equilibrium will be reached only when Ψ_{cell} is zero. Under these conditions, the positive term in Eq. (29), Ψ_{P}, must just balance the negative term; i.e.,

$$\Psi_{\pi} + \Psi_{P} = 0, \text{ and } \pi = P \qquad (30)$$

The water potential equation holds for the cell as a whole, the protoplasm behaving like a semipermeable membrane, so that the entire system functions as an osmometer. Separate potentials can also be identified for the components of the system—the vacuole, the protoplasm and the cell wall. Whenever the water content changes, these components immediately adjust to a new state of equilibrium. That is,

$$\begin{aligned} \Psi_{vacuole} &= \Psi_{\pi \ vac} + \Psi_{P} \\ \Psi_{protoplasm} &= \Psi_{\tau \ ppl} + \Psi_{\pi \ ppl} + \Psi_{P} \\ \Psi_{cell \ wall} &= \Psi_{\tau \ cw} \end{aligned} \qquad (31)$$

At equilibrium, $\Psi_{vacuole} = \Psi_{protoplasm} = \Psi_{cell \ wall}$.

The Single Cell in a Hypertonic Medium; Incipient Plasmolysis. If a cell is placed in a solution more concentrated than that in the vacuole, water is drawn out of the cell. As a result, the volume of the vacuole is reduced and the elastic cell wall is less distended, exerting less pressure on the protoplasts. Finally the volume of the cell

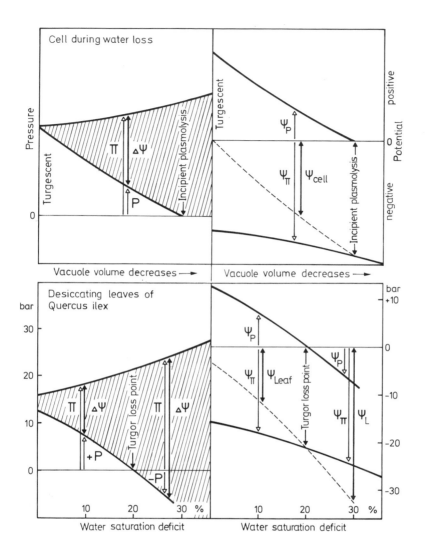

Fig. 86. Osmotic diagrams (left) and water potential diagrams (right) for vacuolated plant cells in a hypertonic medium (above) and for leaves of *Quercus ilex* undergoing desiccation (below). *Osmotic diagram:* As water is withdrawn from the cell the volume of the vacuole decreases, the concentration of the cell sap rises, and the turgor pressure is reduced. The depression of water potential (shaded area) is expressed by the distance between the curve for osmotic pressure of the sap and turgor pressure *P*. When the turgor pressure in wilting scleromorphic leaves falls below zero (loss of turgor), the cell walls can develop tension (negative turgor), the effect of which may be enhanced by the effective suction tension exerted by the cells. *Water potential diagrams:* As water is progressively withdrawn the osmotic potential Ψ_π becomes more negative, the pressure potential Ψ_P falls from positive values to zero (or to negative values in sclerophylls). The water potential of the cells Ψ_C is the sum of the values for Ψ_π and Ψ_P. (After Höfler, 1920; Barrs, 1968; Kyriakopoulos and Larcher, 1975)

shrinks to a limiting value, beyond which the cell wall can no longer follow. At this point the protoplast begins to pull away from the cell wall; this stage is called *incipient plasmolysis*. In the state of incipient plasmolysis, the osmotic pressure is unopposed by the tension of the cell wall:

$$P = 0, \text{ so that } \Psi_{cell} = \Psi_\pi \tag{32}$$

Tissues in Aqueous Media. When cells are not isolated, but form a tissue, their expansion is facilitated or opposed by that of their neighbors; i.e., P in the equation for a given cell is modified by external factors. If the adjacent cells are pushing against the cell considered, P is increased and equilibrium is reached when the water content is still low (and the volume of the vacuole is relatively small). This is important in tissues with delicate walls, for under such conditions they can maintain suitable turgidity without excessive filling with water. If, on the other hand, the cell walls have fused to form a rigid surface, one would expect that they would not readily follow the shrinking protoplasts. Under these conditions P ceases to counteract Ψ_{cell} when only a moderate amount of water has been withdrawn.

Cells of Land Plants in Their Natural Surroundings. The cells of land plants lose water by evaporation from the surfaces of the cell walls. As the cell wall dries out, water flows into it from the protoplasm and, in homoiohydric plants, is replaced by water from the vacuole; the cells begin to wilt. But plasmolysis does not occur, since the cell walls are impermeable to air. The walls follow the protoplasts; they are drawn inward and can even fold up in the process. At the same time tension forces may develop (negative turgor; cf. Fig. 86), which in scleromorphic leaves can enhance considerably the effective suction exerted by the protoplasts. The ecological advantage of this for the sclerophylls is evident; the effective suction—the crucial determinant of the water supply to the leaf—can reach high values without an excessive rise in the osmotic pressure.

Water Translocation in Plant Tissues

The existence of water-potential differences within the cell and in the tissues of vascular plants brings about movement of water in the direction of the lower water potential. Cells can thus replace water lost by evaporation and compensate for local water deficiencies. Within a tissue, water moves between adjacent cells down its concentration gradient—again by osmotic translocation from cells with higher water potential to those in which the water potential is lower. Moreover, water moves from more to less moist cell walls as through along a wick, following the matrix potential gradient. In contrast to most thallophytes, as well as to the textbook model of a single cell in a fluid external medium, the single cells of the homoiohydric vascular plants do not act independently of one another in the management of water. Because of the continual movement of water in the fused cell walls and between the protoplasts, the cell associations constituting individual tissues and organs form coherent functional systems, linked to one another by way of the water-conducting systems and subordinated to the water-balance requirements of the plant as a whole.

Absorption, Transpiration and Water Balance in the Plant

The shoot systems of land plants stand in the air, and steadily lose water that must be replaced from the soil. Transpiration, water uptake and conduction of water from the roots to the transpiring surfaces are inseparably linked processes in water balance. The vapor-pressure deficit of the air (the saturated vapor pressure minus the actual vapor pressure) is the driving force for evaporation, and the water in the soil is the crucial quantity in water supply.

Water Uptake

Plants can take in water over their entire surface, but the greater part of the water supply comes from the soil. In the higher plants this task is delegated to the roots, the specialized organ of absorption. Lower plants are rootless and thus dependent upon direct water uptake *via* organs above ground.

Direct Water Uptake by Thalli and Shoots

Thallophytes imbibe water from damp substrates and from their surfaces after wetting by rain, dew and fog; in the process they swell. Bacteria, lower fungi and some algae and lichens can imbibe water from humid air; they, too, swell as a result. Water uptake by imbibition at first proceeds rapidly and vigorously; it is an exergonic process, releasing heat. As the plants become more swollen, the rate of imbibition slows. Imbibition permits the uptake and storage of large quantities of water. In the maximally swollen state, lichens contain 2—3 times as much water as they do dry matter; in most mosses water exceeds dry matter by a factor of 3—7, and in peat moss and the fruiting bodies of fungi the factor is 10 or more.

Vascular plants are protected against evaporative water loss by a cuticle. To the same degree that the cuticle slows the loss of water through the surface of the shoot, it interferes with the entry of water when the surface is wet. Therefore appreciable direct water intake by shoots occurs only, if ever, at specialized parts of the plant such as hydathodes (water-permeable places in the epidermis) and non-cutinized attachment points of wettable hairs. In epiphytic Bromeliaceae, a significant part of the water supply enters through scales specialized for imbibition.

Water Uptake from the Soil

The Water in the Soil

Water infiltrates the soil following precipitation and gradually percolates deeper, down to the water table. Some of the infiltrating water, however, is retained and stored in the pore spaces of the soil. The relative amounts of water that are retained as *capillary water* in the upper layers of the soil and that sink through them as *gravitational water* depend on the nature of the soil and the distribution of pore sizes within it. Pores up to 60 μm in diameter hold water by capillary action, whereas coarser pores let it seep through.

The Water-Storage Capacity of the Soil. The specific capacity of a soil to retain capillary water is given by the *moisture equivalent*. This is the weight of water the soil retains under a force of 1,000 times gravity, expressed as a percentage of the oven-dried weight of the soil (not of the total wet weight). The water content at saturation of the soils in their natural locations, after the gravitational water has percolated through, is called the *field capacity* and is expressed in the same way. After long periods of rain, and when the snow has just melted in the spring, gravitational water can remain in the upper layers of the soil and thus be available to the roots of plants. In this case, the water content of the soil is greater than the moisture equivalent, by an amount corresponding to the available gravitational water.

Fine-grained soils, as well as soils with a high colloid content and rich in organic substance, store more water than coarse-grained soils. The field capacity thus increases according to the following sequence: sand, loam, clay, muck. A large water-storage capacity is of course advantageous to the plants, enabling them to survive periods of drought.

How Water is Held by the Soil. The water that remains in the soil after the passage of gravitational water is held in the pore space by capillary action; it is also attached to the soil colloids by surface forces and (especially in saline soils) can be bound to ions. Thus the soil water, like the water within a plant, is not entirely free. When the soil is not saturated with water, the forces tending to hold water in the soil are stronger (the water potential becomes increasingly negative). The soil's tendency to hold water can be described by the water potential equation and be given in bar or erg · cm^{-3}. In most soils that part of the total water potential represented by the osmotic pressure of the soil solution is negligibly small, as is P (the hydrostatic pressure of the water in the pore space). The crucial component of the soil's water potential is the *matrix potential* τ—the force with which the water is held by capillary action and adsorption to colloids. The matrix potential is particularly large in narrow pores. As a result, the force with which the soil water is held increases greatly as, during drying out, the large-diameter pores are emptied and capillary water remains only in the finer pores (Fig. 87). In sandy soils with a coarse granular structure the transition is particularly sharp, whereas in loamy and clay soils, in which there is a range of intermediate pore sizes between the average and the smallest, the water potential changes less abruptly.

Water Uptake by Roots

A plant withdraws water from the soil only as long as the water potential of its fine roots is more negative than that of the soil solution. The rate of water uptake is greater, the larger the absorbing surface of the root system and the more readily roots can make contact with water in the soil. According to the formula of W. R. Gardner,

$$W_{abs} = A \cdot \frac{\Psi_{soil} - \Psi_{root}}{\Sigma r} \tag{33}$$

That is, the amount of water absorbed by the roots per unit time (W_{abs}) is proportional to the exchange surface area (A) in the region penetrated by the roots (active root area per unit volume of soil) and the potential difference between root and soil;

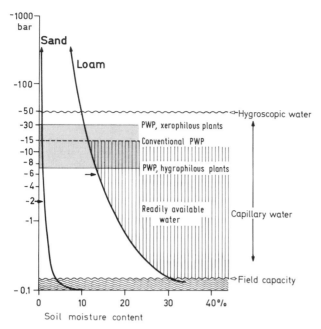

Fig. 87. Diagram of the dependence of the water potential of a sandy soil and a loam soil upon the water content of the soil. *Conventional limiting values*: water is exclusively hygroscopically bound at values of Ψ_{soil} of −50 bar and below; water content at field capacity is considered to correspond to $\Psi_{soil} = −0.15$ bar, and the permanent wilting percentage (*PWP*) to correspond to $\Psi_{soil} = −15$ bar. The *readily available water* depends upon the specific *PWP* of the plants growing on this soil; in plants with high moisture requirement this corresponds to the range −6 to −8 bar, and in plants resistant to drought the *PWP* can correspond to potentials less than −30 bar. The black arrows are referred to in the text. (After Kramer, 1949; Laatsch, 1954; Slatyer, 1967; and data of Gradmann, 1928 and Ellenberg, 1963)

it is inversely proportional to the resistances (*r*) to water transport within the soil and in the passage from soil into plant. The active root surface of herbaceous crops amounts to about 1 cm² · cm⁻³; that of woody plants is of the order of 0.1 cm² · cm⁻³.

The cell-sap concentration in roots is usually enough to give a water potential of only a few bars; this, however, is sufficient to withdraw the greater part of the capillary water from most soils. This effect can be seen in Fig. 87; with Ψ_{root} of only −2 bar the roots withdraw more than $^2/_3$ of the water storable in a sandy soil; a clay soil, which holds water more strongly because of the fineness of its pores, gives up half its capillary water store to roots with a water potential of only −6 bar. To a limited extent, some plants can increase the difference between their water potential and that of the soil solution still further, so as to obtain even more water. Then, however, the force with which water is retained in the soil increases to extremely high values as the water content is reduced. At this stage more water can be taken from the soil only if there is a *replacement* by influx from the regions of the soil in which there are no roots. In sandy soils, with large pores, the columns of water held up by capillarity break under slight tension, so that the supply of water from below is readily interrupted. In clay soils, with very fine capillaries, water is replaced even if the tension is high, but the movement of water is very slow and takes place only over short distances (a few mm to cm). When the "readily available water" in the immediate surroundings of the roots is exhausted, the only possibility remaining to the plant is to follow the water by *root growth* and to enlarge the active surface area of the roots. The root system of a plant moves constantly in the search for water. With advancing desiccation of the soil, parts of a root system can die and dry up, while in other places the root is growing out for many meters and branching profusely. The capacity to do this is especially pronounced in plants growing in dry regions.

Wilting Percentage and the Available Water in the Soil

Eventually, however, a situation may arise in which all sources of water are used up. Then the plants wilt, and do not recover even at night or if one protects them from evaporation (for example, by covering them with a nylon bag). This condition has been called *permanent wilting* by L. J. Briggs and H. L. Shantz. Since the amount of water in the soil at the point of permanent wilting is often expressed as a percentage of the oven-dry weight of the soil, it is called the *permanent wilting percentage* (*PWP*). The *PWP* for herbs is equivalent to a soil-water potential of -7 to -8 bar; those for most agricultural plants lie between -10 and -20 bar, and plants of moderately dry habitats and various woody plants do not wilt until the *PWP* corresponds to -20 to -30 bar.

The difference between the water content at field capacity (W_{FC}) and that at the permanent wilting percentage (W_{PWP}) is a measure of the readily available water (W_{av}) a soil can store.

$$W_{av} = W_{FC} - W_{PWP} \qquad (34)$$

In ecological studies the permanent wilting percentage must be determined for the particular plant species being investigated, but for practical agricultural purposes it has been agreed that the value of W_{PWP} corresponding to a soil-water potential of -15 bar should be used.

Water Uptake and Soil Temperature

The soil temperature influences water uptake by plants in that both the capacity of the roots to absorb and the resistance to movement of water through the soil are temperature-dependent. Plants can extract water from warm soils more readily than from cold soils. At low temperatures the permeability of protoplasm to water falls, and an even more important factor is reduction of the rate of root growth (cf. p. 230), a most critical process in the plant's exploitation of water. In many herbaceous and woody plants, water uptake is considerably reduced at a few degrees above 0° C. In the temperature-dependence of water uptake one can discern adaptation of the plants to the soil temperature prevailing in their habitats (Fig. 88). Species that begin development early are, as a rule, less hampered by low soil temperatures than

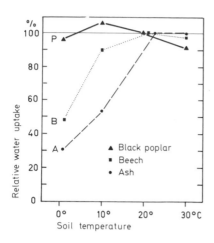

Fig. 88. Temperature dependence of water uptake by the roots of *Populus nigra, Fagus sylvatica* and *Fraxinus excelsior*. (After Döring, 1935)

are species developing later in the season. In beans, tomatoes, cucumbers, squashes and other plants of warmer countries, water uptake is suspended at temperatures just below 5° C, whereas tundra plants and some forest trees can continue to take up water from ground in which the temperature is barely above 0° C or even from ground that is partially frozen. Below −1° C all the capillary water in the soil is frozen (r in Formula 33 is infinitely large), and no water at all can enter the plants.

The Translocation of Water

The Path of Water in the Plant

In the course of *diffusion* of water from cell to cell and within the cell walls, the water moves without obstruction through the root cortex and into the endodermis. In some plants the root cortex is composed of many layers, so that it acts as a reservoir of water to smooth out short-term fluctuations in the rate of influx from the soil. At the endodermis capillary movement of water is prevented by hydrophobic inclusions or lignification of the cell walls (the Casparian strip); the water flowing in is channeled to special cells through which it can pass. In the root stele, water enters a system of long-distance ducts; from this point on, translocation is by *conduction* through the xylem. The parenchyma of the stele is the site of origin of the *root pressure* that results when water (together with nutrient ions) is sent into the duct system at an accelerated rate by means of metabolic energy. The system of conducting vessels is specialized for rapid translocation and distribution, most of the water moving by mass flow through the lumens of the vascular elements. It is possible also to demonstrate conduction via membranes in the long-distance system, but the amounts of water translocated in this way are insignificant. In the leaves the xylem ducts divide into fine branches, and through the tracheids at their tips the water passes to the parenchyma around the veins; from the parenchyma it is displaced to the mesophyll cells, again by diffusion.

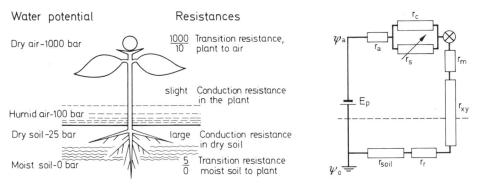

Fig. 89. Water-potential gradients and transport resistances between soil, plant and atmosphere. *Left*: Order-of-magnitude estimates of water potential and resistances. The sharpest potential gradient is that between the shoot surface and dry air. This is also the site of the largest transport resistance; the latter is associated with the high energy requirement for the evaporation of water and with the cuticular resistance to diffusion. (After Kausch, 1955). *Right*: Resistances to water conduction as represented in a circuit diagram. E_p, potential evaporation; Ψ_0, water potential of the liquid phase in the soil; Ψ_a, water potential of the atmosphere; r_{soil}, resistance to diffusion through the soil; r_r, transport resistance in the secondary roots and root cortex; r_{xy}, conduction resistance in the xylem ducts of roots, shoot, leaf petioles and veins; r_m, transport resistance in the mesophyll; r_c, cuticular resistance (very high); r_s, stomatal resistance (variable); r_a, boundary layer resistance; $\otimes$, transition from liquid to vapor phase. (After Cowan, 1965; Boyer, 1974; Kreeb, 1974)

Water-Potential Differences as a Driving Force for Water Conduction

The water flowing through the plant moves upward as a result of the water-potential gradient between atmosphere and soil (Fig. 89). Root pressure contributes somewhat, but is appreciable only when the other components of the potential are insufficient—for example, under conditions of high humidity. Since the shoot is exposed to a much greater vapor-pressure deficit than the subterranean parts of the plant, the various resistances to water conduction within the plant bring about a water-potential gradient from the leaf to the root *via* the xylem ducts; Ψ_{cell} becomes more negative from the bottom of the plant to the top (Fig. 90), as does the pull due to the cohesion of water in the conducting system. If one cuts off a leaf or a twig, this column of water is broken and the water in the ducts is drawn back into the tissues. If the twig is introduced immediately after it is cut off into a pressure chamber like that developed by P. F. Scholander, one can gradually increase the external pressure until the water potential in the cells is compensated and the water moves back into its original positions in the ducts. The extra pressure applied, which can be read off from a manometer, corresponds to the negative pressure previously existing in the twig, a function of the water potential of the cells and the cohesive tension of the water in the xylem ducts of this part of the plant.

Flow Rate of the Transpiration Stream

The *velocity* with which water flows through the conducting system depends upon the fundamental difference in water potential between leaves and roots and upon the

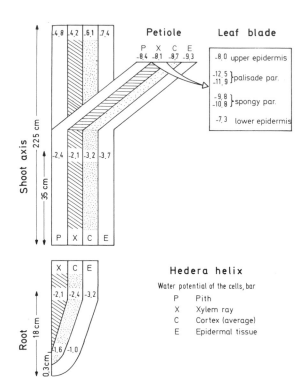

Petiole

Leaf blade

| 4,8 | 4,2 | 6,1 | 7,4 |

P   X   C   E
-8,4 -8,1 -8,7 -9,3

| -8,0 upper epidermis |
| -12,5 / -11,9 } palisade par. |
| -9,8 / -10,8 } spongy par. |
| -7,3 lower epidermis |

Shoot axis

225 cm

| -2,4 | -2,1 | -3,2 | -3,7 |

35 cm

| P | X | C | E |

Hedera helix

Water potential of the cells, bar

P Pith
X Xylem ray
C Cortex (average)
E Epidermal tissue

Root

| X | C | E |
| -2,1 | -2,4 | -3,2 |

18 cm

| -1,6 | -1,0 |

0,3 cm

Fig. 90. Water-potential gradients in an ivy plant. The water-potential depression is always least in the region surrounding the parts of the plant conducting the transpiration stream: in the roots it is greatest in the rhizoderm, where the water is drawn in; above ground, it is greatest in the tissue that is transpiring most vigorously (the epidermis of the shoot axis and the mesophyll of the leaves). Within the plant as a whole, it rises from the bottom to the top and along the petiole to the leaf. (From Ursprung and Blum, 1918)

resistances in the conducting elements. The *amount of water* moved through the system per unit time is greater, the larger the cross-sectional area of the vessels (the "conducting area"). The physiology and ecology of water conduction were emphasized in the research of H. Dixon, J. Böhm, and B. Huber, and it is their work that has laid the foundation for our present-day understanding of these factors in water balance.

The Conducting Area of a shoot axis or the petiole of a leaf is the sum of the cross-sectional areas of all the xylem elements. The conducting area is usually expressed with reference to the mass of the plant parts supplied (i.e., as a *relative* conducting area); for example, in a petiole the conducting area is given as area per unit fresh weight of the leaf, and in a stem as area per total weight of the shoot. The relative conducting area is a measure of the ease with which the individual shoot components of a plant can be supplied with water. It is large in plants that transpire very strongly; some desert plants have relative conducting areas of $2-3$ mm$^2 \cdot$ g^{-1}, and in Mediterranean shrubs, steppe plants and herbs of sunny habitats values between 1 and 2 mm$^2 \cdot$ g^{-1} are found. Most woody plants and sciophytes have conducting areas less than 0.5 mm$^2 \cdot$ g^{-1}. Water plants, as well as succulents, have especially small conducting areas. Within a plant, too, the relative conducting area varies. In trees it increases from the bottom upward, so that the shoots at the treetop are at an advantage. In this way the plant compensates for the longer distance over which water must be conducted.

Table 26. Specific axial conductivity of the xylem, and maximum velocity of the transpiration stream, in various kinds of plants. (From Huber, 1956)

A. Specific conductivity of the woody parts
(ml water passed per hour under 1 bar water pressure, for 1 m sample length and 1 cm^2 cross section)

Conifers	20
Foliage trees, evergreen	14—50
Foliage trees, deciduous	65—130
Lianas	240—1270
Woody roots of deciduous foliage trees	290—5400

B. Maximum velocity of the transpiration stream (m · h^{-1})

Mosses	1.2—2.0
Evergreen conifers	1.2
Larches	1.4
Mediterranean sclerophylls	0.4—1.5
Deciduous diffuse-porous foliage trees	1—6
Ring-porous foliage trees	4—44
Herbs	10—60
Lianas	150

Conduction Resistances. The transpiration stream must overcome a series of resistances as it rises through the plant; the force of gravity, filtration resistances in the transverse walls crossing the conducting elements at certain intervals, and above all the friction in narrow vessels, must be overcome. The specific resistance of conducting elements in the shoot axis (or its inverse, the specific conductivity) is a measure important in characterizing the water translocation system of a plant. The specific conductivity of conifers is half that of evergreen foliage trees, and the latter in turn is half that of deciduous foliage trees (Table 26). The roots, with their large-diameter vessels, conduct water particularly well, as do lianas.

The Velocity of the Transpiration Stream. The maximum possible velocity of the transpiration stream depends upon the structure of the conducting system concerned (particularly as it affects resistance to flow) and varies among different plant parts and different types of plants (Table 26). As long as water uptake by the roots is not hampered, the rate of flow increases with the main driving force, the rate of evaporation (Fig. 91). The transpiration stream adjusts extremely rapidly to the rate of transpiration, reflecting even short-term fluctuations; thus one can use measurements of flow velocity in tree trunks to infer the progress of transpiration in the entire crown of the tree. In larger trees movement of water begins in the morning at the very top of the crown and at the tips of the branches, drawing up the column of water extending to the base of the trunk. Then the transpiration stream begins to flow rapidly, and soon after sunrise it attains its maximum rate. In the evening flow

149

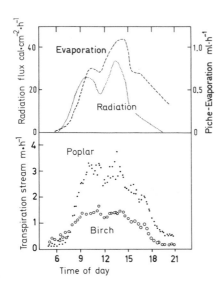

Fig. 91. Daily variation in the velocity of the transpiration stream in poplars and birches. When radiation and low humidity (high evaporation rate as measured by Piche evaporimeter) promote transpiration through leaves, the water is translocated more rapidly through the conducting elements. (After Klemm as cited by Polster, 1967)

becomes slower, but until late at night there can be a slow influx to the trunks, so that their water reserves are replenished.

Water Loss from Plants

Plants lose water through evaporation (transpiration) and occasionally also, to a slight extent, in liquid form (guttation). As far as water balance is concerned, guttation is of little significance, so that in the following discussion, when water loss is mentioned, it is always transpiration that is meant.

Transpiration as a Physical Process

Water evaporates from the entire outer surface of a plant, and from all interior surfaces that come into contact with air. In thallophytes the outer surfaces of the thallus are involved, whereas in vascular plants the corresponding surfaces are the cutinized epidermis (*cuticular transpiration*) and the surfaces of cells bordering on intercellular air space. In the latter case, the water first is converted from the fluid phase to the vapor phase, and then the water vapor escapes through the stomata (*stomatal transpiration*). From the surface of the plant the vapor diffuses into the adjacent layer of air (the boundary layer, cf. p. 28) and thence into the open air. The movement of water vapor from the evaporating surface to the open air is brought about by diffusion, in accordance with Fick's Law (Formula 8, p. 27).

Evaporation from Moist Surfaces

As a physical process, transpiration by plants follows the rules governing the evaporation of water from moist surfaces. An exposed water surface gives off more water vapor per unit time and area, the sharper the gradient of vapor pressure in the air. A vapor-pressure gradient arises when the water-vapor content of the air at the evaporating surface is greater than that at some distance from this surface. This is always the case when the evaporating surface is adequately supplied with water and is warmer than the air. Strong irradiation warms the surface and thus leads to a sharper vapor pressure gradient and to more rapid evaporation. Evaporation under conditions of unlimited water supply and unrestricted diffusion of the water vapor is called *potential evaporation* (E_p) and is directly proportional to the *evaporative power* of the air. The *actual evaporation* from moist surfaces (soil, plant cell walls) is usually less than the potential evaporation, since there is almost never a complete replacement of the lost water.

Stomatal Transpiration

Stomatal transpiration can be formulated as a diffusion process:

$$E_s = \frac{C_i - C_a}{r_a + r_s} \tag{35}$$

That is, the stomatal transpiration E_s (in g $H_2O \cdot cm^{-2} \cdot s^{-1}$) is proportional to the difference between the water-vapor content in g $H_2O \cdot cm^{-3}$ within the leaf (C_i) and in the atmosphere (C_a), and it is limited by the sum of the diffusion resistances—that is, the stomatal resistance r_s and the boundary-layer resistance r_a.

The *boundary-layer resistance* (units of reciprocal velocity or s·cm^{-1}) to water-vapor diffusion, like that to CO_2 exchange, is very dependent upon air movement; with a wind velocity of 0.1 m · s^{-1} it is in the region of 1–3 s · cm^{-1}, and when the wind velocity is 10 m · s^{-1} it falls to 0.1–0.3 s · cm^{-1}.

Because of the unfavorable ratio of diffusion path length to cross-sectional area, the very small stomatal slits constitute the chief obstacle to the diffusion of water vapor. The minimal *stomatal diffusion resistance* (with the stomata wide open) depends upon the size and structural peculiarities of the stomata and upon their arrangement and density. Comparative measurements of stomatal diffusion resistance and anatomical investigations of leaf structure provide an objective basis for determining the maximal transpiration rate in different species. For example, P. G. Jarvis and P. Holmgren could show that the minimal diffusion resistance for water vapor is lowest in herbs (0.3–2 s · cm^{-1}) and distinctly higher in deciduous foliage trees (2–10 s · cm^{-1}); the values for pine needles are 5–10 times as great as those for the leaves of foliage trees. Conifer needles often have porous plugs of wax in the substomatal chambers, which act as additional obstacles to transpiration and reduce water loss to $^1/_3$ of that possible in their absence. Moreover, there are differences within these groups of plant types; birch and poplar have a lower r_s than maple, and that of the oak is higher.

The foregoing results are in good agreement with those obtained by measurements of transpiration (Table 27). The maximum transpiration of leaves (mass of H_2O per unit area per unit time) is greatest in some swamp and floating plants such as *Potamogeton* and *Alisma*; among land plants the strongest transpiration is found in herbs growing in sunny habitats, whereas sciophytic herbs under the same condi-

Table 27. Transpiration rates of leaves of various plants, with evaporative power of the air 0.4 ml $H_2O \cdot h^{-1}$ (as measured with a Piche evaporimeter). All data given in mg $H_2O \cdot dm_2^{-2}$ (surface area on both sides of the leaf) $\cdot h^{-1}$. (From Pisek and Cartellieri, 1931, 1932, 1933, and Pisek and Berger, 1938)

Plant	Total transpiration with open stomata	Cuticular transpiration after stomatal closing	Cuticular transpiration as % of total
Herbaceous heliophytes			
Coronilla varia	2000	190	*9.5%*
Stachys recta	1800	180	*10%*
Oxytropis pilosa	1700	100	*6%*
Herbaceous sciophytes			
Pulmonaria officinalis	1000	250	*25%*
Impatiens noli tangere	750	240	*32%*
Asarum europaeum	700	80	*11.5%*
Oxalis acetosella	400	50	*12.5%*
Trees			
Betula pendula	780	95	*12%*
Fagus sylvatica	420	90	*21%*
Picea abies	480	15	*3%*
Pinus sylvestris	540	13	*2.5%*
Evergreen Ericaceae			
Rhododendron ferrugineum	600	60	*10%*
Arctostaphylos uva ursi	580	45	*8%*

tions lose only half as much water. Trees and dwarf shrubs, in turn, lose still less than the sciophytic herbs. Among trees, heliophytes transpire more strongly than sciophytes; poplar and birch, with their low stomatal diffusion resistance, lose more water than beech under comparable conditions. Flowers have very few stomata and thus lose relatively little water.

Cuticular Transpiration

Cuticular transpiration refers to the diffusion of water molecules through the cutinized layers of the outer wall of epidermis and through the cuticle. One can thus consider cuticular transpiration as diffusion through a hydrophobic medium. The *cuticular diffusion resistance* is usually very high; it varies in different species with arrangement, density and number of the cutin and wax lamellae embedded in the outer wall of epidermis, as well as with the thickness of the cuticle. In hygromorphic leaves the cuticular diffusion resistance is $20-100$ s $\cdot$ cm^{-1}, but in xeromorphic leaves and needles with massive protection against transpiration, it can reach values near 400 s $\cdot$ cm^{-1}. When the outer epidermis dries out and shrinks, the hydrophobic layers are brought closer together, and as a result the cuticular diffusion resistance can double. At low temperatures, too, the cuticular diffusion resistance increases. The protection against transpiration provided by cutin is very effective. Even in plants of shady and damp habitats, cuticular transpiration amounts to less than 10% of the evaporation at a free water surface; in sclerophylls and evergreen conifer needles it is reduced to 0.5%, and in cacti to as little as 0.05%, of the potential evaporation. Intermediates between these extremes are represented by herbaceous heliophytes, trees and dwarf shrubs (Table 27).

The loss of water through the surfaces of suberized shoot axes is comparable in magnitude to cuticular transpiration; the amount of such loss, in any case very small, depends upon the species-specific structure of the periderm, the permeability of the lenticels, and the presence or absence of cracks in the bark. For this reason poplars, oaks, maples and pines give off more water than spruce, beech and birch trees with their smoother, denser bark.

The Dependence of Transpiration upon External Factors

Overall transpiration (including both stomatal and cuticular transpiration) depends upon external factors to the extent that these factors affect the process of evaporation. Thus the rate of transpiration rises most markedly with increased irradiation and the associated warming, and it also rises when the air is relatively dry or in motion. Wind carries away the humid film of air just next to the epidermis and brings new, unsaturated air up to the evaporating surface; the vapor pressure gradient, and thus the rate of transpiration, is increased.

Physiological Control of Transpiration

Transpiration is determined by the physical conditions affecting evaporation only as long as the degree of opening of the stomata does not change—that is, as long as the stomata remain open to a fixed degree or remain firmly closed. Under these conditions, the amount of water lost is proportional to the evaporative power of the air. By changing the degree of stomatal opening, however, the plant can regulate its transpiration in accordance with the requirements of its water balance.

Relative Transpiration

One can detect the intervention of *physiological* regulatory mechanisms by comparing the rate of transpiration of a plant to the simultaneous rate of evaporation from a moist surface. To determine the latter, plant ecologists use conventional, standardized evaporation devices (called evaporimeters or atmometers). One such device employs disks of thick filter paper that are kept moist; the amount of water lost by evaporation is measured (in mg or ml) per unit time and per unit area (or per disk). Another device, the Livingston atmometer, permits evaporation from a measured reservoir of water through a porous pottery bulb; evaporation is expressed as ml H$_2$O lost per unit time and is converted to standard units according to the calibration of the instrument. Atmometer measurements are not sufficiently precise for the physical analysis of evaporation phenomena, but they are useful in comparative studies, particularly in the determination of *relative transpiration* as suggested by B. E. Livingston:

$$E_{rel} = \frac{E}{E_p} \tag{36}$$

where E is the transpiration rate and E_p is the potential evaporation at the same time and place. Changes in transpiration brought about by regulation of stomatal opening cause an increase in this ratio when the stomata open and a decrease when they close.

The relative transpiration of a plant naturally changes with time; a graphic presentation of the course of such change is given by E/E_p *diagrams* (as devised by A. Pisek and E. Cartellieri). Fig. 92 shows such a diagram and explains its interpretation.

Stomatal Regulation of Transpiration

Changes of stomatal pore area are elicited by different external and internal factors (cf. p. 30). With respect to water balance, the most important of these is the closing of the stomata when the water supply is poor. Under strong illumination—and cor-

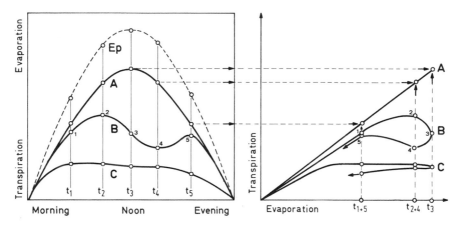

Fig. 92. Schematic diagram of the *daily variation* in transpiration by a plant well supplied with water (unrestricted transpiration, A), a tree with water supply limited at noon (*B*), and a grass (*C*) under continual regulation of stomatal transpiration, as compared with evaporation from a moist disk of filter paper (potential evaporation, E_p). Right: the *transpiration/evaporation* (E/E_p) *diagram* corresponding to the figure on the left. For each data point representing evaporation, the corresponding (simultaneous) rate of transpiration is noted. When transpiration is unrestricted, the line joining these points is straight—transpiration is proportional to evaporation. Restrictions on the rate of transpiration are reflected in a departure from proportionality; in (*B*) the rate of transpiration falls from time t_2 to time t_3 although the rate of evaporation is rising during that interval, so that the diagram forms a loop. Any departure from a diagonal in the E/E_p diagram indicates stomatal regulation of transpiration. In the literature, the scale for potential evaporation is commonly reduced by a factor of 10 as compared with that for transpiration

respondingly high photosynthetic activity—the CO_2 control system initially counteracts the tendency to close, but if water balance is severely impaired the H_2O control system prevails and the guard cells collapse.

The response threshold, the rapidity, and the effectiveness of stomatal regulation vary among species and with the degree of adaptation to the habitat. Trees and herbaceous sciophytes narrow their stomatal openings even when there is only a slight water deficiency, and briskly complete the process of closing. Herbs of sunny habitats restrict their stomatal transpiration only under much drier conditions, and even then closing is slow (cf. Fig. 105). There are many small differences in the way stomata behave in different plant species and even in different individuals of the same species. In fact, even within a single plant, leaves vary appreciably in this respect, depending on their form and their position on the shoot.

Once the stomata have closed, cuticular transpiration determines the rate of water loss. The specific *effectiveness of stomatal closing* can be expressed by the ratio between unrestricted total transpiration and cuticular transpiration (Table 27). In soft-leaved plants, cuticular transpiration accounts for an average of $^1/_3$ to $^1/_{10}$ of the total, whereas in sclerophylls it contributes only $^1/_{20}$ to $^1/_{30}$.

Transpiration and CO_2 Exchange

As already emphasized, transpiration and CO_2 acquisition by a plant are linked by the stomata, through which both water vapor and CO_2 diffuse. In order to take up CO_2 the plant must give off water, and when water loss must be reduced the influx of CO_2 is reduced as well.

The rates of diffusion of the two substances, however, are not identical. The CO_2 concentration difference between the outside air and the chloroplasts is much less than the water-vapor pressure difference between the interior of the leaf and the atmosphere, as long as the outside air is not saturated with water. At a temperature of 20° C and a relative humidity of 50%, the water-vapor gradient is about 20 times as sharp as the CO_2 gradient. For this reason alone, the evaporation of water proceeds much more rapidly than the uptake of CO_2. Moreover, the water molecules, being smaller, diffuse 1.5 times as fast as the larger CO_2 molecules, given identical gradients. There is also a fundamental difference with respect to the diffusion pathways. The route is longer for CO_2, which must enter the chloroplasts, and there is an additional obstacle in that CO_2 movement in solution occurs exceedingly slowly. The ratio of transpiration to photosynthesis (E/F) is therefore always changed whenever any of the factors affecting diffusion is altered. When the stomata are open, CO_2 uptake is limited more severely than transpiration by the diffusion resistances inside the leaf (above all by the carboxylation resistance). When the stomata are closed, CO_2 uptake is blocked, but water continues to escape through the cuticle so that the E/F ratio, for practical purposes, approaches infinity. The most favorable compromise between water consumption and CO_2 uptake is achieved when the stomata are partially open. This is evident not only in the light-dependence relation but also in the behavior during progressive desiccation; as the plant begins to dry out, the E/F ratio is lowest when both exchange processes are somewhat restricted (Fig. 93). In the natural habitat, the expenditure of water necessary to obtain the

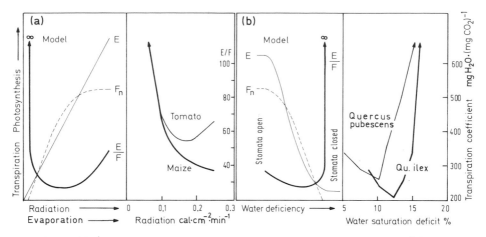

Fig. 93 a and b. Transpiration, photosynthesis and transpiration coefficient (E/F) as a function of insolation, evaporative power of the air and water supply. (a) As the incident radiation increases, the stomata open and the evaporative power of the air becomes greater. Transpiration steadily increases, but photosynthesis increases only up to the saturation level, which is reached earlier in C_3 plants (tomato) than in C_4 plants (maize). At intensities of illumination a little below the light-saturation region for photosynthesis, the transpiration coefficient falls to a minimum. The "model" curve (valid for C_3 plants) is taken from De Wit (1958), and the transpiration coefficients for tomato and maize from Barrs (1966). (b) Limitation of E and F with increased water deficiency and the associated closing of stomata. The transpiration coefficient is particularly low when the stomatal pores have already narrowed somewhat, at the beginning of the closure brought about by desiccation. In this respect there are clear differences between species (compare the curves for the deciduous oak *Quercus pubescens* and the evergreen oak *Quercus ilex*). When the stomata are closed the transpiration coefficient approaches infinity, because photosynthesis is entirely suppressed while the cuticular component of transpiration continues to operate. (The model curve is original, the data taken from Larcher, 1960)

CO_2 for photosynthesis is least in the early hours of the morning, since photosynthesis gets off to a more rapid start than does transpiration. As warming by the sun progresses and air humidity decreases, the rate of water loss rises more rapidly than that of CO_2 uptake, and the E/F ratio increases.

The chief problem associated with gas exchange, as formulated by O. Stocker, is that of "tacking adroitly between thirst and starvation". Some plants can do this better than others and are thus more successful competitors in dry habitats. For the management of plantations in agriculture and forestry, where the target is the greatest possible productivity, it is important to know the relationship between water consumption and productivity. This relationship can be expressed by the *transpiration coefficient* or by the *productivity of transpiration*; the first of these is given in terms of the amount of water used by a plant or a stand of plants during the growing season per unit weight of dry matter produced, and the second refers to the amount of dry matter that can be produced per liter of evaporated water. The transpiration coefficient varies among different species and is very dependent upon the conditions

Table 28. Average water consumption during the production of dry matter (g transpired water per g dry matter produced). (From Stocker, 1929; Polster, 1967; Black, 1971)

Herbaceous plants		Trees	
C$_4$ plants		**Deciduous trees**	
Maize	370	Oak	340
Millet	300	Birch	320
Amaranthus	300	Beech	170
Portulaca	280		
C$_3$ plants		**Conifers**	
Rice	680	Pine	300
Rye	630	Larch	260
Oats	580	Spruce	230
Wheat	540	Douglas fir	170
Barley	520		
Alfalfa	840		
Bean	700		
Crimson clover	640		
Potato	640		
Sunflower	600		
Watermelon	580		
Cotton	570		

in the habitat and the degree of openness of the stand. Some relevant data are given in Table 28. Such measurements enable an exact determination of the amount of irrigation required for cultivated plants in dry regions.

The Water Balance of a Plant

The water balance of a plant is given by the difference between the rates of water intake and water loss:

$$\text{Water balance} = \text{water absorption} - \text{transpiration} \qquad (37)$$

Here transpiration is viewed as a measure of the expenditure of water and not as a physical process; therefore it is expressed in terms of the amount of water lost per unit mass (usually fresh weight) and not per unit area. It would be better, however, to compare absorption and transpiration with the water content of the plant, expressing the *water turnover* as mg water evaporated per g water content. The *water turnover rate* gives the percentage of the water present in the plant (in the leaf) which is lost in a given period of time (minute, hour, day) and must be replaced if the water balance is to be maintained.

Occasionally the relationship between water loss and water replacement is expressed as a ratio, but this is a less relevant measure than the difference between the two decisive quantities determining water balance.

A satisfactory water balance can be maintained only if the rates of uptake, conduction and loss of water are suitably adjusted. The balance becomes negative as soon as the supply of water no longer meets the transpiration requirements. If the stomata narrow as a result of this deficit, so that transpiration is decreased while uptake continues as before, the balance is restored after a transient overshoot to positive values. Thus the water balance of a plant oscillates continually between positive and negative deviations. It is useful to distinguish the short-term oscillations from the long-term disturbances of this equilibrium. Short-term fluctuations reflect the interplay of the various water-regulatory mechanisms, particularly the changes in stomatal aperture. There are more marked departures from equilibrium in the course of a day, particularly in the alternation between day and night. In the daytime, in a natural habitat, the water balance is almost always negative; the water content of the plant is not restored until evening or during the night (and then only if there are sufficient water reserves in the soil). During dry periods the water content is not entirely restored overnight, so that a deficit accumulates from day to day, until the next rainfall; similarly, there are seasonal fluctuations in water balance.

Determination of Water Balance

Water balance can be *computed directly* from quantitative determinations of water uptake and transpiration. In the laboratory, this can be done with a "potetometer". In the field, however, the methodological difficulties in determining water uptake are so great that one generally makes an *indirect estimate* of the water balance through its effect upon water content or water potential of the plant. A negative balance always eventually produces a decrease in turgidity and water potential of the tissues. These changes appear first in the leaves, which are the site of most intensive evaporation and moreover are the furthest removed from the roots.

Change in Water Content as an Indicator of Water Balance. A water deficit can be demonstrated by repeated measurements of the water content of leaves and other parts of the shoot. The actual water content at any given time (W_{act}) must be given with respect to a standard measure—for example, the water of the leaves under conditions of saturation (W_s). The water content at any particular time of observation can be expressed either as a percentage of the water content at saturation (the *relative turgidity* of P. E. Weatherley) or as a *water saturation deficit* (as suggested by O. Stocker). The water saturation deficit (*WSD*, in %) indicates how much water a tissue lacks as compared with complete saturation.

$$WSD = \frac{W_s - W_{act}}{W_s} \cdot 100 \qquad (38)$$

The water saturation deficit increases as a result of maintained negative water balance, and the relative turgidity under the same conditions decreases.

Changes in Water Potential. Fluctuations in water content necessarily affect the concentration of the cell sap and the water potential of the cells. The *osmotic pres-*

sure, as a component of the water potential of the cell, provides an indication of changes in the water balance. The osmotic pressure rises as long as water balance is negative. The absolute values of π differ according to plant species, form, state of development and the tissue concerned. The range within which osmotic pressure fluctuates also varies in these respects. The optimum value (π_{opt}) is considered to be the osmotic pressure in a state of optimum water content; the osmotic maximum (π_{max}) is the value found in nature under conditions of extreme drought. There are plants which can support great osmotic stress without damage (*euryhydric* species). In contrast, *stenohydric* species suffer impairment of their vital functions with even slight increases in osmotic pressure. To obtain a picture of the water balance in plants of different climatic regions and habitats, one may, as suggested by H. Walter, summarize the full range of variation of π values in the graphic form of "osmotic spectra". In such a diagram (Fig. 94) one can see that the plant species fall into ecologically distinct groups.

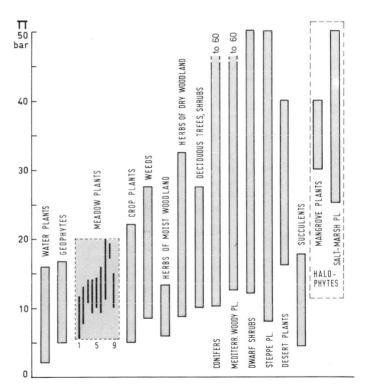

Fig. 94. Ranges for the values of osmotic pressure of leaves of ecologically different types of plants (the *osmotic spectrum*). The sub-ranges (black bars) shown for the meadow plants illustrate how to interpret the osmotic range given for each plant group; that is, it is derived from the difference between the lowest and the highest osmotic pressures found among all plants studied in the ecological group. *1 Polygonum bistorta; 2 Taraxacum officinale; 3 Galium mollugo* and *Campanula rotundifolia; 4 Achillea millefolium; 5 Tragopogon pratensis; 6 Poa pratensis; 7 Melandrium album; 8 Cynodon dactylon* and *Lolium perenne; 9 Arrhenatherum elatius.* (After Walter, 1960)

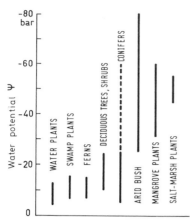

Fig. 95. Water potentials of leaves and twigs of ecologically diverse plant types (the xylem sap tension measured in a pressure chamber under strong daytime insolation). (After Scholander *et al.*, 1965)

The most sensitive indicator of water balance is the *water potential* Ψ of the leaves and twigs. The loss of turgor associated with water deficit lowers directly the water potential of the cells. Ψ also becomes more negative as a direct result of an inadequate water supply. The absolute magnitude and range of variation of Ψ in different species can also (as for π in Fig. 94) be arranged to show a sequence of ecological groups (Fig. 95).

Basic Types of Water Balance

In every climatic zone and habitat, plants with quite diverse water relationships grow side by side. They can be classified in two fundamental categories with respect to

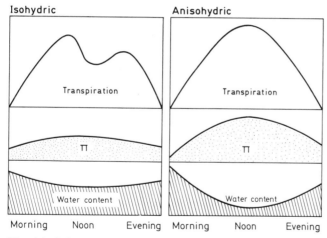

Fig. 96. Schematic diagram of the two basic categories of water-balance mechanism as proposed by Berger-Landefeldt. The *isohydric* type (left) avoids pronounced fluctuations in water content and osmotic pressure during the day by stomatal regulation of the rate of water loss. The *anisohydric* type does not restrict transpiration until it has become very dry; the latter mechanism allows the water balance to stay negative for extended periods, as is reflected in the wide fluctuations of osmotic pressure and water content in the course of a day. (From Stocker, 1956)

water balance, the hydrostable (isohydric) and hydrolabile (anisohydric) types (Fig. 96).

Hydrostable species can, to a great extent, maintain a favorable water content throughout the day, the water balance remaining near zero; their stomata respond with great sensitivity to lack of water, and their root systems as a rule are extensive and efficient. A further factor stabilizing protoplasmic water content is the presence of water reserves in storage organs, roots, the wood and bark of the stems and the leaves. Hydrostable plants include trees, some grasses, shade plants and succulents. Often hydrostable species are stenohydric, but they are not necessarily so.

Hydrolabile species are in danger of large losses of water and marked increases in cell-sap concentration. Their protoplasm must be able to tolerate these rapid and extensive fluctuations in water potential. Hydrolabile vascular plants are therefore always euryhydric. Many herbs of sunny habitats are hydrolabile, as are the poikilohydric plants.

Water Balance in Different Plant Types

Trees. In mature trees, with their extensive evaporating surfaces and the long distances water must travel from roots to leaves, it would be especially serious if appreciable water deficits were allowed to develop; incipient water loss must be dealt with from its first appearance. The entire crown of a tree can transpire without limit only on overcast or very cloudy days, and when an adequate water supply is assured. On fine days, at least around noon, trees always encounter difficulties in keeping up with the rate of water loss, so that the guard cells—which in most trees respond even to very small water saturation deficits—temporarily restrict transpiration. Later in the day, when the water content has been restored, the stomata open again and the rate of transpiration increases (Figs. 92B and 97). The *noon depression of transpiration* is characteristic of trees on clear days. If the water deficit of the tree is not alleviated sufficiently during this period, the afternoon rise in rate of transpiration may be less pronounced or completely absent. Water-conservation

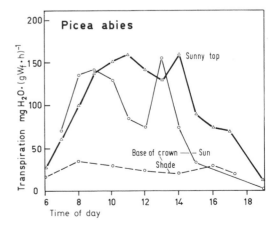

Fig. 97. Daily fluctuations in the transpiration of spruce shoots on a sunny August day which had been preceded by dry weather. With an inadequate supply of water, the shoots in the shade at the base of the crown first reduce their water loss, then the twigs in the sun at the lower margin of the crown, and finally the shoots in the sunny top of the crown. (After Pisek and Tranquillini, 1951)

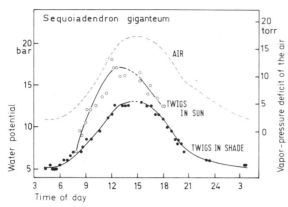

Fig. 98. Daily changes in water potential in twigs of the giant sequoia at a height of 7 m (measured with a pressure chamber). As the vapor pressure deficit of the air increases in the morning, the water potential in the twigs becomes more negative, more rapidly in twigs located in the sun than in the shade. With the decline in vapor pressure deficit during afternoon and evening, water balance is restored and the water potential changes in the opposite direction, becoming least negative only a short time before sunrise. (From Richter *et al.*, 1972)

measures are not taken in all parts of the crown at once, but occur in a stepwise manner and in an orderly sequence: the first and most prominent reduction of transpiration occurs in the shady parts of the crown, then in the base of the crown, and finally even the leaves of the tree top limit their evaporation. Correspondingly, the rise in water potential in the sunny tops of trees is sharper than in the shaded parts, and these topmost leaves are therefore preferentially replenished with water (Fig. 98). The range of daily fluctuation of osmotic pressure is very small in trees, rarely exceeding an amplitude of 3 bar; this means that in spite of the increase in the forces drawing water upward during the course of a day the protoplasm itself is not subjected to any great fluctuations in water potential. Not all trees are hydrostable to the same extent. Trees characterized by good maintenance of water content are conifers, sciophytic species and some heliophytic species such as oaks. There are also trees—for example, species of ash—that do not behave so conservatively; during dry spells the leaves wither prematurely.

Herbaceous Dicotyledons. Among the herbaceous plants one finds all intermediate forms between hydrostability and extreme hydrolability. This diversity of types is particularly evident in sunny habitats with a tendency to dryness of the soil; examples are given in Fig. 99.

When there is a fairly adequate water supply, some of the dicots transpire without restriction throughout the day, at a rate determined by the evaporative power of the air (Fig. 92A). Others behave more like trees, depressing the rate of transpiration at midday or even limiting the period of maximal stomatal opening to the early morning and late evening. The extravagant transpiration displayed in the first group can be afforded only by plant species which either can draw upon abundant water reserves through an extensive root system or are able to endure a high degree of

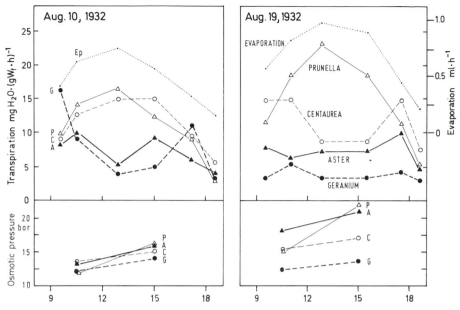

Fig. 99. Daily changes in transpiration and osmotic pressure in various plants from a dry, hot habitat at the beginning of a dry period (left) and after 9 days of drought (right). *Prunella grandiflora* is a shallow-rooted plant that barely reduces its rate of transpiration and thus is subject to marked excursions of its water balance. *Centaurea scabiosa* has moderately deep roots and reduces stomatal transpiration only when it becomes difficult to maintain the water supply; *Aster amellus* also has moderately deep roots but responds to slight changes in water balance and restricts transpiration at an early stage; *Geranium sanguineum* is a very shallow-rooted plant which at the slightest shift of its water balance drastically narrows its stomatal apertures, thus largely avoiding an increase in osmotic pressure even after several days of dryness. (After Müller-Stoll, 1935)

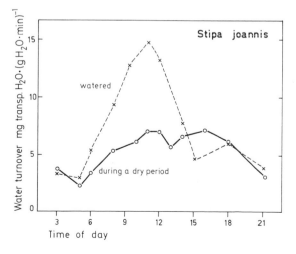

Fig. 100. Daily changes in transpiration of the steppe grass *Stipa joannis* during a dry period in July, as compared with the transpiration of artificially watered plants. Here transpiration is given in terms of water turnover (mg transpired water per g water contained in the leaves). (After Rychnovská, 1965)

desiccation. In the latter case, the osmotic pressure of the cell sap can rise by 3—6 bar during the course of a day. Such species are found frequently among the herbaceous heliophytes (cf. Fig. 105).

Grasses. These, too, comprise both hydrostable and hydrolabile species. The stomata of hydrostable grasses respond with extreme sensitivity to the first sign of negative water balance. The stomata therefore close gradually during the morning so that often no sharp depression is observed around midday (Figs. 92C and 100). As the leaves age—at the beginning of the dry period in savanna grasses—the stomata lose their mobility, and the grasses gradually lose control of the water balance. Even though the soil is dry, the plants continue to transpire without restraint until the leaves wither.

Water Balance during Drought

Restriction of Water Consumption during Summer Dry Periods

When no rain falls for weeks or months at a time, the water reserves in the soil are depleted and the water content of plants progressively diminishes; the plants tend more and more to limit their water consumption by opening their stomata less, and for shorter periods. In Fig. 101 this trend is illustrated: at first transpiration is reduced during the hottest hours of the day, then the afternoon resumption of transpiration is omitted, and finally the stomata open only in the morning. Eventually, while their water content is still adequate, the plants transpire only through the cuticle. Examples of the effect of several weeks of dryness in summer upon plants of xerothermic habitats in central Europe can be found in Figs. 99 and 102. Under such conditions economical use of water is discernible even in those herbaceous dicots that previously had transpired maximally the whole day long (Fig. 99). At most, certain deeply rooted plants may continue to transpire freely for some time

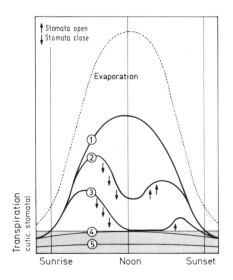

Fig. 101. Diagram of daily changes in transpiration as it becomes progressively more difficult (curves 1—5) to maintain the water supply. The arrows indicate the stomatal movements elicited by changes in the water balance. The stippled area shows the range in which transpiration is exclusively cuticular. *1* unrestricted transpiration; *2* limitation of transpiration at noon as the stomata begin to close; *3* full closure of the stomata at midday; *4* complete cessation of stomatal transpiration by permanent closure of the stomata (only cuticular transpiration continues); *5* considerably reduced cuticular transpiration as a result of membrane shrinkage. (After Stocker, 1956)

164

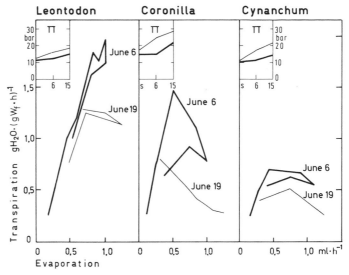

Fig. 102. Transpiration as a function of evaporation (E/E_p diagrams), and range of fluctuation of osmotic pressure during the day, for different species in a xerothermic habitat at the beginning of a dry period (June 6, heavy lines) and after two weeks of drought (June 19, light lines). The osmotic pressure is indicated for water-saturated leaves (s) as well as for leaves collected in the morning (0600) and during the greatest departure from equilibrium in the afternoon (1500). The looped curves are to be interpreted as discussed in Fig. 92. *Leontodon incanus* is a deep-rooted plant which maintains a stable water balance despite a high rate of transpiration, *Coronilla varia* is a shallow-rooted plant that readily reduces its water consumption but still cannot avoid a marked rise in osmotic pressure; *Cynanchum vincetoxicum* has both shallow roots and such sensitive stomatal regulation that there is only a moderate rise in osmotic pressure even in the dry period. (After Pisek and Cartellieri, 1931)

(*Leontodon* in Fig. 102). In those species showing more judicious water management even before the onset of drought, the stomata are kept closed almost continuously after the drought has lasted for several weeks.

Woody plants, as a rule, elaborate a root system extending to considerable depths and are thus not forced to restrict their transpiration as radically as are herbs. Nevertheless, the leaves of trees and shrubs in regions of periodic dryness give off during the summer drought only $1/10 - 1/3$ of the water lost during the wet season (Table 29).

Winter Drought Effects

Water deficiencies can also occur in winter and can even cause injury if replenishment of the water supply is prevented by prolonged frost. In regions with long cold winters (mountains, the arctic and subarctic), the soil remains frozen near the surface for months, and the water ducts in shoots and roots become filled with ice (the water in the tracheary elements is frozen at temperatures below $-2°$ C). The twigs not protected by snow, however, continue to lose water. The water balance becomes

Table 29. Total daily transpiration of woody plants in periodically dry regions, comparing the rate under conditions of adequate water supply with that under the conditions of evaporation prevailing in the habitat during the dry season. [From Rouschal, 1938 (a); Grieve and Hellmuth, 1968 (b); Stocker, 1970 (c); Hellmuth, 1971 (d)]

Plants	Total daily transpiration $(g\ H_2O \cdot dm_2^{-2} \cdot d^{-1})$		Level to which transpiration is reduced	Author
	Rainy season	Dry season		
Mediterranean maquis				
Quercus ilex	9.5	1.0	*11%*	a
Laurus nobilis	3.3	0.5	*15%*	a
Olea europaea	10.0	2.8	*28%*	a
SW-Australian sclerophyllous woodland				
Acacia acuminata	12.4[a]	2.9[a]	*23%*	d
Acacia craspedocarpa	18.6[a]	3.9[a]	*21%*	d
Banksia menziesii	7.4	0.6	*8%*	b
Australian bush				
Hakea preisii	19.8	13.4	*68%*	d
Eremophila miniata	33.3	17.9	*54%*	d
Desert plants				
Nitraria retusa	19.1	15.0	*78%*	c

[a] $g\ H_2O \cdot (gW_d \cdot d)^{-1}$

negative, and the water content deteriorates progressively. The problem is greatest in late winter, when the soil has not yet thawed but the sun increasingly warms the branches and stimulates transpiration (Fig. 103).

Drought Resistance

Drought resistance is the capacity of a plant to withstand periods of dryness. This capacity is a complex characteristic. The prospects for survival of a plant under extreme stress due to drought are better, the longer a dangerous decrease in the water potential of the protoplasm can be delayed (the avoidance of desiccation) and the more the protoplasm can dry out without becoming damaged (the capacity to tolerate desiccation). In the terminology of J. Levitt,

$$\text{Drought resistance} = \text{drought avoidance} + \text{drought tolerance} \tag{39}$$

Among vascular plants drought tolerance is strictly limited, so that differences in the

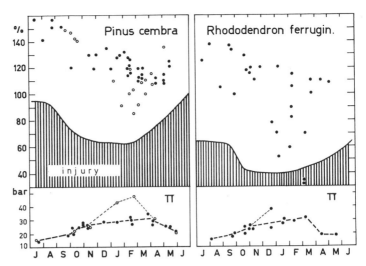

Fig. 103. Water content (as % of dry weight), osmotic pressure and drought resistance of pine needles and the leaves of *Rhododendron ferrugineum* at the alpine treeline. The circles in the pine diagram denote small trees (up to 2 m high), and the dots refer to mature trees. Pine needles and *Rhododendron* leaves acquire an enhanced protoplasmic resistance to desiccation at the beginning of their winter dormant period, and do not lose it until the beginning of the growing season in the spring. In late winter, when the ground is still frozen but the sun warms more strongly the branches not protected by the snow cover, the water content of the *Rhododendron* and of small pine trees falls, approaching and sometimes passing the limit for safety of the plant. Because of the water reserves in their trunks, larger trees are not as greatly endangered as are young trees and shrubs. (After Larcher, 1972)

drought resistance of different species are chiefly ascribable to the varying effectiveness of the mechanisms for drought avoidance.

Drought Avoidance

All the physiological and morphological properties of a plant that permit it to maintain a favorable water content for the longest possible time, in the face of atmospheric and soil dryness, contribute to drought avoidance. These include

1. improved water uptake from the soil through decrease of the water potential and extension of the root system,
2. reduced water loss by timely closure of the stomata, effective prevention of evaporation by the cuticle, and reduction of the transpiring surface area, and
3. water storage and an increased capacity for water conduction.

By growing rapidly into deeper regions of the soil or through cracks in stone, the roots reach horizons more likely to retain water. The seedlings of woody plants in dry regions send out taproots ten times as long as the shoot, while the grasses in such places elaborate a dense root network and also send their threadlike roots to depths measured in meters. The ratio between the masses of shoot and root is shifted further in favor of the roots, the more the exposure to drought (cf. Table 8).

167

The relationship between transpiring surface and water stored in the tissues is expressed by the *specific leaf area* and the *degree of succulence* of the leaves or shoots:

$$\text{Specific leaf area} = \frac{\text{Surface area (dm}^2)}{\text{Fresh weight of leaf (g)}} \tag{40}$$

$$\text{Degree of succulence} = \frac{\text{Water content at saturation (g)}}{\text{Surface area (dm}^2)} \tag{41}$$

Alternatively, one may compare leaf area with the dry weight of the entire plant: this is the *leaf area ratio*.

Since the rate of water evaporation increases with the transpiring area, a limited leaf area (or high degree of succulence) conserves the water stores of a leaf. Leaves developed under conditions of poor water supply are, as a rule, correspondingly smaller and more strongly subdivided and have a smaller specific leaf area. The leaf venation is narrower, the stomata are closer together, and the outer wall of epidermis, with the cuticle, is thickened.

A particularly effective way of *reducing the transpiring surface* of a plant is the partial or complete abscission of the leaves. A number of woody plants in dry regions shed their leaves regularly during the drought season. Through peridermal transpiration, the trunks and branches of large trees lose only $^1/_{300}$–$^1/_{3000}$ of the amount of water evaporated from the leaves when the water supply is good. Species of *Ephedra, Sarothamnus, Cytisus, Spartium, Retama,* and desert shrubs like *Cilla macroptera* and *Zollikoferia arborescens* cast off their leaves as required, reducing the surface area to $^1/_3$–$^1/_5$ of its usual value. The folding and rolling up of grass blades also results in restriction of transpiration (in *Stipa tenacissima* transpiration is reduced to as little as 40% of normal).

Drought Tolerance

Drought tolerance refers to the species-specific adaptable capacity of protoplasm to endure severe desiccation. This resistance to desiccation is measured by equilibrating parts of plants or pieces of tissue (sections or cut-out disks), unprotected against transpiration, with air of known humidity. The lowest relative humidity (or the corresponding water-potential values) at which the cells are just capable of survival (the "critical limit") or are damaged by a certain amount (e.g., 50% injury is the drought lethality, DL_{50}) is used as the measure of tolerance. In the literature one also finds tolerance given in terms of water content—for example, critical and sublethal (5–10% injury) water content or water deficit. Such data, if carefully interpreted, are informative with respect to the detection of adaptive and seasonal changes in drought tolerance, but they should not be used for comparisons of different species since their absolute values are influenced by the anatomical peculiarities of the sample investigated. For example, the fraction of the measured weight associated with mechanical strengthening elements can alter the figure. The drought tolerance of plant protoplasm varies over a wide range (Table 30).

Drought-Tolerant Species. All orders of thallophytes contain species capable of tolerating complete desiccation. Many representatives of the bacteria, cyanophytes

Table 30. Drought resistance of plant cells after 12—48 hrs in vapor chambers with different relative humidities. [From Iljin, 1927, 1930 (a); Biebl, 1938 (b); Höfler, 1942, 1950 (c); Abel, 1956 (d); Sullivan and Levitt, 1959 (e); Parker, 1968 (f)]

Plant		Tolerated without injury		Moderate injury		Author
		% *RH*	bar	% *RH*	bar	
Marine algae						
Sublittoral algae		99—97	14—41			b
Algae of the ebb line		95—86	69—204			b
Intertidal algae		86—83	204—252			b
Liverworts						
Hygrophytes	usually	95—90	69—141	92—90	112—141	c
Mesophytes	usually	92—50	112—933	90—36	141—1400	c
Xerophytes	usually	(36)—0	(1400)—∞	0	∞	c
Mosses						
Water mosses and hygrophytes		95—90	69—141			d
Mesophytes	usually	90—50	141—933			d
	extreme	10	3000			d
Xerophytes	usually	5— 0	400—∞			d
Tracheophytes (tissue sections)						
Leaf epidermis				96—92	55—112	a
Mesophyll		96	55	95—90	69—141	a, e
Root cortex				97—95	41—69	f

and lichens can withstand a state of dryness (in equilibrium with the surrounding air) for months and even years, resuming metabolic activity as soon as they are moistened again (Table 31). Some of them can survive weeks in absolutely dry air over concentrated sulfuric acid or phosphorus pentoxide. Other completely drought-tolerant plants include some fungus mycelia, various mosses, and some pteridophytes. Among the flowering plants there are only a few drought-tolerant species—e.g., *Ramonda serbica* and some species of *Haberlea* (both Gesneriaceae) in the Balkan region, and species of Scrophylariaceae, Velloziceae, Myrothamnaceae, Cyperaceae, and Poaceae in South Africa.

Drought-Sensitive Species. The protoplasm of most plants is extraordinarily sensitive to loss of water and dies after even a slight reduction in water content. The cells of homoiohydrous vascular plants perish if they are exposed without protection for a few hours to a relative humidity between 92 and 96% (which in a closed system would be in equilibrium with a solution having $\pi = 55—110$ bar); roots are even more sensitive. Plants may become "hardened" to dry conditions, but this shifts the tolerance limit only slightly (3—4% *RH*). During growth periods the cells are especially sensitive to desiccation; during dormant periods they are somewhat more

Table 31. Drought tolerance of various lichens from southwestern Germany. (From Lange, 1953)

Species	Origin	When kept in an air-dry condition	
		weeks without injury	weeks until death
Lobaria pulmonaria	Forest	8	38
Usnea florida	Forest	8	54
Cetraria islandica	Nardetum	38	54
Peltigera canina	Beechwood floor	38	54
Peltigera canina	Mesobrometum	38	62
Umbilicaria cylindrica	Rock (1450 m)	54	69
Umbilicaria pustulata	Rock (1390 m)	62	94

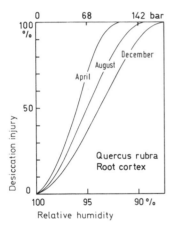

Fig. 104. Drought tolerance of the cortical parenchyma of oak roots at different seasons. Drought resistance was determined by placing sections in chambers with graded relative humidity until equilibrium was obtained. The measure of desiccation injury was the percentage of dead cells in each sample as compared with a control at 100% relative humidity. The relative humidity (or the corresponding absolute value of Ψ_{cell}, in bar) at which the desiccation injury amounts to 50% is the measure of *drought tolerance*. (After Parker, 1968)

resistant (Fig. 104; cf. also Fig. 103). Among the thallophytes the most sensitive are the planktonic algae and those seaweeds attached so far below the surface that they are normally always covered by water; mosses are also sensitive and are found only where the prevailing humidity is continuously high. Plants restricted to permanently moist habitats are called *hygrophytes* and those that can exist in dry places, *xerophytes*. All species not specialized toward either of these two extremes are *mesophytes*. The last category also includes the algae of the tidal zones, which are regularly left dry at ebb tide and lose considerable amounts of water before they are again submerged. Most mosses, too, are mesophytes; often they are adapted to a narrow range of water potential, so that the various species (there are differences even between species of a single genus) are of value as indicators of the humidity of a habitat. It should be said, of course, that "hardening" can considerably increase the resistance of mosses to desiccation.

Drought can be so extreme that plants are no longer capable of extracting any water at all from the soil. Drought resistance after complete cessation of water replacement is termed *specific survival time* (*Überdauerungsvermögen*). This is a measure of the degree to which a plant species can conserve the water stored in its shoots. According to A. Pisek, specific survival time is computed from the cuticular transpiration (E_c) and the available water (W_{av}) in the plant—that consumed between the time of stomatal closure and the appearance of the first signs of desiccation injury.

$$\text{Specific survival time} = \frac{W_{av}}{E_c} \tag{42}$$

Survival time is measured in hours or days and indicates how long after stomatal closure the leaves of a plant species can remain undamaged without a supply of water, for a given evaporative power of the air. The amount of available water is greater if the stomata close as soon as there is a slight water deficit, if the tissues can store an abundance of water, and if desiccation injury is incurred only with large water deficits (Fig. 105). Survival time thus depends both on the effectiveness of

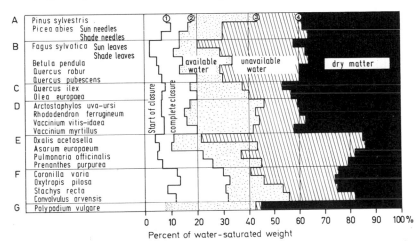

Fig. 105. Water content of the leaves of various plants under conditions of saturation, and availability of this water store. *1* water deficit as compared with the saturated weight at the *onset of stomatal closure; 2* saturation deficit after *complete stomatal closure; 3* saturation deficit at the appearance of the first signs of desiccation injury. The remaining water (3 → 4) is necessary for life and can be diminished only at the risk of severe damage ("*unavailable water*"). The "*available water*" (which can be used after stomatal closure until the water content is reduced to the point of incipient damage; dotted area) represents an emergency reserve for dry periods. The species are grouped as follows: *A* conifers; *B* deciduous foliage trees; *C* evergreen foliage trees; *D* dwarf shrubs; *E* herbaceous sciophytes; *F* herbaceous heliophytes; *G* poikilohydric fern. The various groups differ in the threshold of their response and in the time-course of stomatal closure, as well as in the ratio of available to unavailable water. (After Pisek and Winkler, 1953, with addition of data from Larcher, 1960; Boyer, Y., 1968)

Table 32. Specific survival time of leaves cut from various plants; the evaporative power of the air was 0.4 ml $H_2O \cdot h^{-1}$ as measured by Piche evaporimeter. (Computed from data given in Pisek and Berger, 1938; Pisek and Winkler, 1953; Larcher, 1960)

Plant	Specific survival time		Water consumption after stomatal closure (cuticular transpiration)		Water available from stomatal closure to appearance of damage (cf. Fig. 105)	Specific leaf area (leaf surface on both sides per saturation weight)	
	Hrs	Rel. (Pinus = 100)	mg H_2O $\cdot$ dm$_2^{-2}$ $\cdot$ h^{-1}	Rel. (Pinus = 1)	mg H_2O $\cdot$ dm$_2^{-2}$	dm$_2^{-2}$ $\cdot$ g^{-1}	Rel. (Pinus = 1)
Sclerophyllous plants							
Pinus sylvestris	50	100	13	1	660	0.4	1
Picea abies	22	44	15	1.2	340	0.4	1
Quercus ilex	17	35	7	0.6	120	0.6	1.5
Rhododendron ferrugineum	9.5	19	60	4.6	560	0.5	1.2
Soft-leaved plants							
Asarum europaeum	3.5	7	80	6.2	270	1.2	3
Convolvulus arvensis	2.5	5	120	9.2	305	0.8	2
Fagus sylvatica							
shade leaves	2.1	4	55	4.2	115	1.6	4
sun leaves	0.7	1.4	90	7	60	1.4	3.5
Quercus pubescens	2.2	4.4	40	3	90	1.1	2.9
Quercus robur	1.5	3	110	8.5	160	0.9	2.3
Oxalis acetosella	1.2	2.4	50	3.8	60	1.8	4.5
Coronilla varia	0.8	1.6	190	14.5	155	0.7	1.8
Pulmonaria officinalis	0.5	1	250	19	130	1.2	3
Impatiens noli tangere	0.4	0.8	240	18.5	90	2.2	5.5
Succulents							
Sedum maximum	20	~40	170	13	~3300	0.12	0.03
Opuntia camanchica	1000	~2000	1.8	0.14	~20000	0.026	0.0065

drought avoidance and on the degree of drought tolerance. From Table 32 it can be seen that of the factors prolonging life, the cuticular surface protection varies most among different species. Only the succulents have a remarkable specific survival time, attributable primarily to their ability to store water.

Relative Drought Index

Plants differ with respect to survival capacity; the extent to which individual plants, *in their habitats*, suffer from dry conditions depends not only upon their drought resistance but also on the conditions prevailing in the habitat. Both of these factors are comprised in the *relative drought index (Trockenheitsbeanspruchung)*. As defined by K. Höfler, H. Migsch, and W. Rottenburg, this index expresses as a percentage the relative magnitudes of the actual water saturation deficit (WSD_{act}) measured at a given time in the habitat and the critical water saturation deficit (WSD_{crit}) for the species concerned:

$$\text{Relative drought index} = \frac{WSD_{act}}{WSD_{crit}} \cdot 100 \tag{43}$$

Table 33 includes a selection of examples showing that plants of the same species can be endangered by drought to different degrees in different habitats and that, in contrast, different plant species growing close together can be affected to a very different extent by drought, depending upon their constitution.

Specific survival time and relative drought index are both useful in characterizing the prospects of survival of a species under conditions of water deficiency. *Specific survival time* is particularly suited to describe the probability that a plant will last through a season of dryness, for the effects of drought depend upon its *duration*. By reference to the *relative drought index* one can obtain additional information about *spatial differences* in severity of drought if several individuals of the same species, in different locations, are compared with one another. This measure reflects the degree to which the water stores in the soil and the local conditions for evaporation can vary; drought is thus appreciated as a habitat characteristic with a crucial influence upon the distribution of species.

Water Economy in Plant Communities

The Water Balance of Stands of Plants

The state of water balance in a stand of plants, and in the soil penetrated by its roots, can be expressed by the *water balance equation*, a formula similar in structure to those for the carbon balance (17) and mineral balance (25) of ecosystems. All the quantities involved are given as precipitation equivalents in mm H_2O (i.e., liters per m^2 of ground):

$$Pr = \Delta W + L_E + L_O \tag{44}$$

Table 33. Relative drought indices for various plants. [From measurements by Oppenheimer, 1932 (a); Rouschal, 1938 (b); Höfler *et al.*, 1941 (c); Arvidsson, 1951 (d); Bornkamm, 1958 (e); Larcher, 1960 (f); Rychnovska, unpubl. (g)]

Plant	Place	Average maximal water saturation deficit in the natural habitat in % of *WSD* at first appearance of damage	in % of *WSD* at 5—10% damage	Author
Evergreen trees and maquis shrubs of the Mediterranean region				
Olea europaea	Israel	83%		a
Pinus halepensis	Israel	80%		a
Laurus nobilis	Israel	80%		a
Laurus nobilis	Istria	40%		b
Viburnum tinus	Istria	70%		b
Quercus ilex	Istria	63%		b
Quercus ilex	Garda Lake region	47%		f
Subshrubs of the Mediterranean garrigues				
Rosmarinus officinalis	Israel	(105%)[a]		a
Cistus villosus	Istria	90%		b
Deciduous trees and shrubs				
Ficus carica	Israel	64%		a
Populus tremula	Southern Sweden	38%		d
Quercus robur	Southern Sweden	15%		d
Fraxinus excelsior	Southern Sweden	10%		d
Sambucus nigra	Southern Sweden	28% (52%)[a]		d
Cornus sanguinea	Southern Sweden	27%		d
Herbaceous undergrowth of forests				
Fragaria vesca	Southern Sweden	25% (50%)[a]		d
Oxalis acetosella	Southern Sweden	14%		d
Pulmonaria officinalis	Southern Sweden	6%		d

[a] in parenthesis = extreme values

174

Table 33 (continued)

Plant	Place	Average maximal water saturation deficit in the natural habitat		Author
		in % of WSD at first appearance of damage	in % of WSD at 5—10% damage	
Herbaceous dicots of xerothermic habitats				
Anthyllis vulneraria	Austria	96%		c
Anthyllis vulneraria	Germany		56%	e
Aster linosyris	Austria	93%		c
Helianthemum canum	Austria	88%		c
Hieracium pilosella	Austria	85%		c
Hieracium pilosella	Germany		86%	e
Cynanchum vincetoxicum	Austria	78%		c
Cynanchum vincetoxicum	Southern Sweden		61% (75%)[a]	d
Anemone pulsatilla	Austria	50%		c
Bupleurum falcatum	Austria	43%		c
Lotus corniculatus	Germany		43%	e
Steppe grasses				
Stipa capillata	Moravia		80% (108%)[a]	g
Stipa stenophylla	Moravia		76% (88%)[a]	g
Stipa joannis	Moravia		55% (92%)[a]	g
Stipa pulcherrima	Moravia		50% (87%)[a]	g
Meadow grasses				
Brachypodium pinnatum	Germany		74%	e
Festuca pseudovina	Germany		38%	e
Bromus erectus	Germany		28%	e
Avena pratensis	Germany		19%	e
Cultivated plants				
Capsicum annuum	Austria	71%		c
Lupinus albus	Austria	61%		c
Solanum lycopersicum	Austria	56%		c
Cucurbita pepo	Austria	56%		c
Helianthus annuus	Austria	52%		c
Soja hispida	Austria	46%		c
Daucus carota	Austria	44%		c

[a] in parenthesis = extreme values

175

Under the simplifying assumption that the only input to the plant cover is precipitation, the situation is such that when averaged over years and decades, the water intake (the mean *total precipitation Pr*) is accounted for by evaporation from plants and soil (loss by *evapotranspiration*, L_E) and by *runoff and percolation* through the soil (L_O). Over shorter periods, however, the *water stores in the ecosystem* increase (positive ΔW) or decrease (negative ΔW), since occasionally more rain water falls than evaporates and drains off, or at times not enough falls to meet the requirements of the plants. In the hydrology literature, ΔW is considered to include only the water reserves in the soil—i.e. the amount of capillary water and available gravitational water. In the temperate zone, the amount of water contained in the soil is greatest after the snow melts in the spring; during the summer the water content decreases steadily, despite occasional replenishment by precipitation, until a minimum is reached in late summer. In dry regions, the water reserves are filled up during the rainy season, a process requiring weeks, until even the deeper layers of the soil are thoroughly wet. From an ecological point of view, ΔW must also include the water stored in the phytomass and in the layer of litter. More than $^3/_4$ of the green plant mass, and half the mass of wood, is water; the water content of the layer of vegetation fluctuates in the course of the day and the year (being maximal when the plants bear leaves; cf. Fig. 106), and the total increases as the biomass grows.

Available Precipitation

The amount of precipitation available to the plants for maintenance of their water balance is that which reaches and penetrates the ground. In dense stands of plants not all of the precipitation (*Pr*) actually reaches the ground; rather, the amount entering the ground is only that fraction that falls through holes in the plant canopy, or drops off the leaves or runs down the stems. The resulting local unevenness in the distribution of precipitation is particularly pronounced in woodland, where the ground becomes wetter under holes in the crowns of the trees, under the outer regions of the crown (as a result of dripping), and above all in the vicinity of the trunk. The amount of stem flow is greater, the steeper the angle of the branches and the smoother the bark; at the bases of beech trees more than 1.5 times as much water infiltrates the soil as in the open.

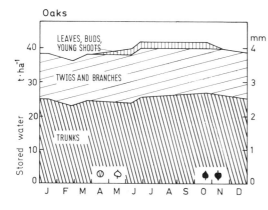

Fig. 106. Water storage in trunks, branches, twigs, young shoots, buds and leaves of a stand of oaks. Phenological symbols from left to right: bud swelling, leafing out, yellowing of leaves, leaf fall. (After Schnock, 1972)

Some of the precipitated water adheres to the plant surfaces and evaporates. Trees in particular intercept large quantities of water in this way. In a forest, therefore, one can count as "intake" only the net precipitation (Pr_n) that falls through, drips from, or runs off the trees. Only a minuscule fraction of the water wetting the trees is taken in directly through the leaves and bark. By far the greater part of the intercepted water evaporates, so that for practical purposes all the water retained by vegetation is treated as a loss (the loss by interception L_I). Thus,

$$Pr_n = Pr - L_I \qquad (45)$$

Net precipitation = total precipitation − interception

Strictly speaking, then, the water balance equation for a stand of plants should read

$$Pr - L_I = \Delta W + L_E + L_O \qquad (46)$$

The loss by interception depends upon the composition and density of the plant cover and upon the meteorological conditions prevailing during the precipitation. Dense crowns of trees with small, easily wettable leaves or needles retain more precipitation than open crowns with large, smooth leaves; the degree to which leafing-out has progressed, of course, is also important. On the average, the loss by interception in coniferous forests amounts to 30% and in leafy forests, to 20% of the total precipitation. As weather conditions change, the fraction intercepted varies over a wide range according to amount and type of precipitation (rain, dew or snow), according to the prevailing temperature and according to the wind. In general more precipitation is intercepted, the finer the drops and the smaller the total amount of water that falls (Fig. 107). Clearly, a certain amount of water is required to wet

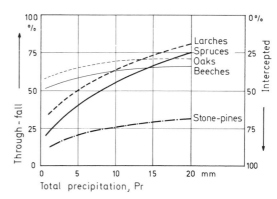

Fig. 107. Relative amount of water falling through (left ordinate) and intercepted (right ordinate) by the canopies of stands of trees under different amounts of precipitation; foliage trees are represented by thin lines, and conifers, by thick lines. The units on the ordinates are % of total precipitation. The crown of the stone-pine is very dense and even under heavy rainfall intercepts a great deal of the precipitated water. (After Hoppe as cited by Walter, 1960, Ovington as cited by Geiger, 1967, and Turner as cited by Aulitzky, 1968)

Table 34. Transpiration by stands of plants. (From the data of numerous authors, taken from summaries in Pisek and Cartellieri, 1941; Stocker, 1956, 1972; Rutter, 1968; Zelniker, 1968; Mitscherlich, 1971; Rychnovska et al., 1972)

Stand	Region	Transpiration mm per year	mm per day	Precipitation mm per year	Transpiration in % of total precipitation
Forests and stands of trees					
Eucalyptus plantation	S. Africa	1200		760	160
Tree plantations	Java	2300–3000		4200	55–72
Tree plantations	Brazil	600		1400	43
Evergreen rainforest	Kenya	1570		1950	80
Bamboo forest	Kenya	1150		2160	53
Mixed forest	Europe, Japan, USA	500–860		1000–1600	50–54
Coniferous forest	Central Europe	580		1250	46
	Northern taiga	290		525	55
Forested steppe	Russia	(110) 200–400		400–500	(25) 50–80
Maquis	Israel	500		650	77
Chaparral	California	400–500		500–600	80–83
Heath and tundra					
Ericaceous heath under pine wood	Russia	115–130		500	24–26
Alpine dwarf-shrub heath	Central Alps	100–200	1.6–3.1	870	11–23
Lichen tundra with moss cover	Siberia	80–100		500	16–20
Grassland					
Sedge and reed	Germany	1300–1600		800	160–190
Wet meadow	Austria	1160	15.5	860	135
Wet meadow	S. Moravia	–	8.3	560	–
Grain fields	Germany	ca. 400		800	50

Pasture (clover and turf)	Germany	ca. 400		800	50
False-oat meadow	Austria	320	4.3	860	37
Dry grassland	Austria	200	2.6	860	30
Dry grassland	S. Moravia	–	2.0	560	–
Steppe	Bechuanaland	200		430	46
Steppe	S. Moravia	–	0.5–1	560	–
Alpine grassland	Austria	50	1.1	1100	5
Open vegetation					
Desert	Mauretania	–	0.03–0.2		–
Desert	Israel	–	0.01–0.4		–
Alpine cushion and rosette plants on calcareous ground	Northern Alps	18	0.4	1100	1.7
Upper alpine cushion and rosette plants on silicaceous ground	Central Alps	11	0.3	> 1100	<1.0

the plants thoroughly, and it is only thereafter that the water can begin to drip off the leaves and twigs. The water needed for wetting the plants sets a threshold for net precipitation; in conifers it is distinctly higher than in foliage trees (3–6 mm as opposed to 0.5–2 mm); heath and pasture land retain 1–2 mm, and a cover of peat moss about 15 mm, before the water begins to reach the ground.

Evaporation from a Stand of Plants

Table 34 summarizes the water consumption of stands of plants in different climatic regions. Most of the values are computed from measurements of the transpiration of single plants or parts of plants, though some are estimates based on large-scale determinations of water balance in natural terrain. The present state of our knowledge of water consumption by plant cover is incomplete and needs, in particular, confirmation by the simultaneous application of several methods to the same stand of plants. The most important of such procedures are the weighing of large bricks of soil covered with vegetation (by an automatic-weighing lysimeter), calculations based on ecophysiological measurements of transpiration of single plants, and computation of the water turnover in homogeneous stands of plants, based on their energy balance and their water-vapor exchange (geophysical methods).

The procedure most frequently used in the past has been the extrapolation from individual plants to the stand. Especially in the case of *stands of herbs*, one can thus obtain useful estimates. With this method, the estimated water consumption of a stand increases about in proportion to the mass of the shoots (Fig. 108), independent of the species composition of the stand. The reason for this is that under comparable weather conditions, total daily transpiration differs by at most a factor of 6 among even widely different species of herbs, whereas the fresh weights of individual plant communities can differ by a factor of 100. Similar results are observed when one computes the average transpiration of the crowns of trees or *stands of trees*. In this case, extrapolation from the transpiration of twigs to that of the whole crown gives only a rough indication, since the different parts of the crown are nonuniform and

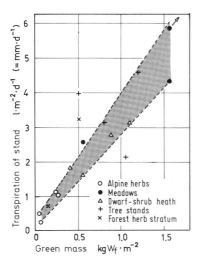

Fig. 108. Average daily water consumption of various stands of plants as a function of the plant mass. With rare exceptions, the transpiration of the stand depends less upon the specific rate of transpiration of the different plants than upon the mass of the plants above ground. (After Pisek and Cartellieri, 1941; Polster, 1950, 1967)

overlap so as to shield one another with respect to the movement of water vapor. The crowns of trees, and to an even greater extent the forest, create a climate for themselves, so that the local conditions for evaporation depart considerably from those found in the open. Nevertheless, one can obtain some idea of the water turnover of mature trees: a birch, for example, loses about 100 l of water by evaporation on a sunny summer day, and during the entire growing season a beech loses about 9,000 l.

The next step in the calculation proceeds from the daily totals to average values for the growing season or for the whole year. An example of the course of transpiration in a stand of reedlike grass is given in Fig. 109. This plant community is always adequately supplied with water, but the amount of transpiration fluctuates considerably from day to day, following the changing conditions for evaporation. Such fluctuations are much larger if there is a water deficiency so that stomatal regulation comes into play, which must be expected to occur among plants on land. In this case one must know not only the average daily transpiration but also the minimal water turnover when the water supply is poor. These limiting values are especially important if one wishes to determine the annual water consumption of the vegetation in countries with dry seasons. In the temperate zone transpiration falls to minimal values in late autumn and during the winter months. This is chiefly due to the lowered evaporative power of the air, but an additional factor is that the evergreen plants do not open their stomata during the period of winter dormancy, and in particular when the temperature is below freezing.

Transpiration of a stand depends primarily upon the transpiring *plant mass* and the local *water supply* of the plants (Table 34). Under similar climatic conditions, forests (probably because of their greater mass) transpire appreciably more than grasslands, and these in turn transpire more than heath. The greatest water turnover is always found in stands of plants growing in marshy habitats (or in any sort of wet land) or

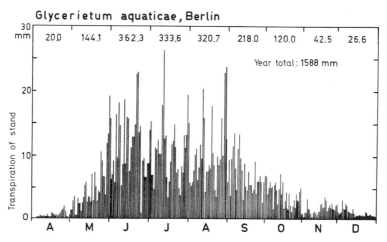

Fig. 109. Daily and monthly totals for transpiration of a stand of reeds during one growing season. The daily water consumption fluctuates according to the weather conditions. (After Kiendl, 1953)

in places where the plants have ready access to groundwater. These plants transpire much more water than is brought in by precipitation, and sometimes even more than would evaporate from an open water surface. The most limited water turnover is exhibited by stands in dry regions. As water becomes less plentiful, the stands are more open; but the separation of individual plants is increased only above the ground—in the soil, the root systems extend over wide areas. In a forested steppe, the distance between the trees is determined not by meeting of the crowns but rather by the radii of the root systems of the individual trees. For regions with dry seasons, there is little point in giving annual data; more useful is an average value for the rainy period and the minimal value for the dry period.

Quantitative data for the average and minimal water consumption of stands of plants are valuable as a basis for decision-making in forestry and landscape management, as well as for irrigation projects, if they are interpreted in connection with the expected precipitation. For example, it has been possible to calculate that open stands of trees can exist only where they receive at least 110 mm precipitation in a year (10—12 mm per month). The water-balance equation (Formula 42) indicates just where forestation becomes uneconomical. On the other hand, the enormous transpiration of trees that tap the groundwater (e.g., poplars and eucalyptus) can be employed to reduce the water table and to increase air humidity.

Runoff and Percolation

Not all the water reaching the ground is available for evapotranspiration. Some of it runs off the surface of the ground, and another fraction percolates into the deeper layers of the ground and joins the groundwater, which in humid regions is a subterranean form of drainage, appearing on the surface here and there as springs, maintaining connections with the rivers, and eventually falling to sea level. The surface runoff is relatively easy to measure, particularly when circumscribed catchment areas of a river are studied. Subterranean drainage, however, must be estimated indirectly.

The amount of water drained away depends primarily on the slope of the terrain and the type of vegetation. In Table 35 precipitation, evapotranspiration and drainage are compared for sections of terrain each covered by uniform vegetation. In dry regions the water is soaked up by the soil and little drains off. In regions of heavy precipitation, it is important for water retention that the falling water penetrate rapidly the layers covering the ground and the soil itself. In loose forest soils with a thick layer of litter, the water infiltrates the soil most rapidly. There is no superficial runoff from woods on flat land; all the water percolates and flows away as groundwater. Precipitated water penetrates much more slowly into the soils of meadows with a dense mat of roots and into pasture soils, compacted by the weight of grazing animals. Frozen soil, too, slows the percolation of melted snow, which accumulates in depressions or, if the land is sloped, runs off.

Where there is a *steep slope*, more than half the precipitation flows off over the surface, and where the precipitation is heavy and woodland is limited (e.g., mountain meadows) as much as $^2/_3$—$^3/_4$ is lost in this way. Because of the marked effects of gravity in mountainous regions, water balance here is of considerable importance to

Table 35. Water balance in extensive stands of plants. (From summaries in Duvigneaud, 1967; Stanhill, 1970; Mitscherlich, 1971; Grin, 1972)

Stand	Region	Precipitation mm · yr^{-1}	Evapotranspiration L_E in % of precipitation	Drainage L_O (surface and groundwater) in % of precipitation
Wooded areas				
Tropical rain forest	Congo basin	1900	73	27
Savanna	Congo basin	1250	82	18
Eucalyptus plantation	Israel	640	87	13
Mixed forest	Central Europe	600	67	33
	NE Asia	700	72	28
Coniferous forest	S. Germany	730	60	40
	Russia	800	65	35
Mountain forest	Switzerland	1640	52	48
	German central mountains	1000	43	57
	Southern Andes	2000	25	75
Grassland and agrarian regions				
Mountain pasture	Switzerland	1720	38	62
Alfalfa fields	S. Germany	715	62	38
Buckwheat fields	Russia	600	68	32
Steppe	Russia	500	95	5

man; white-water conditions, rockslides, erosion and avalanches are intimately related to the drainage and water-storage capacity of the soil and its plant cover.

Additional Water Supplies to the Plant Cover

Because precipitation is not the exclusive source of water for plants, the simplified water-balance equation must be extended by terms that take into account the additional sources of water in the surroundings of a stand. Chief among these are the *stream of groundwater*, which can be utilized by deep roots, the provision of *surface water* from watercourses, and artificial *irrigation*. In dry regions a permanent cover of vegetation can become established only by drawing upon the groundwater, and even in humid climates trees soak up a great deal of groundwater. Fig. 110 shows how the water table beneath a forest gradually sinks, reaching its lowest point in autumn, and is refilled over the winter; in the absence of trees (for example, after clear-cutting) the ground-water level remains nearly constant throughout the year. By exploiting deep-lying stores of water, plants accelerate the circulation of water in the biosphere, for they pump directly back into the atmosphere water which otherwise would have to flow to the sea and return *via* a much longer pathway.

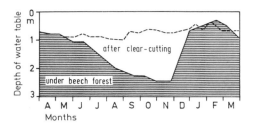

Fig. 110. Variations in water table under a beech forest during the year, before and after clear-cutting. Extraction of water by the trees lowers the water table during the growing season, whereas after the area is cleared the groundwater remains at about the same level throughout the year. (After Holstener-Jørgensen as cited by Mitscherlich, 1971)

Regional Water Turnover

The order of magnitude of the individual factors contributing to water balance can be estimated for large areas, if all available data are evaluated critically. In Fig. 111 such estimates for the water balance of West Germany are presented.

In this region, 57% of the area is used for agriculture, 28% is wooded and the remaining area is built up, is used for industry, or amounts to wasteland. Precipitation is derived from evaporation by the ocean (41%) and by evapotranspiration from land areas (59%). The latter quantity comprises evaporation from water surfaces, the ground, and the surfaces of plants, as well as transpiration—by far the largest

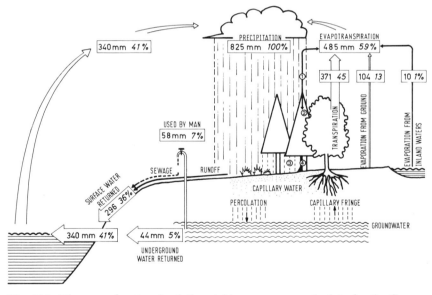

Fig. 111. Diagram of water circulation, taking as an example data for the German Federal Republic, as given in Clodius (1970) and R. Keller (1971) for the period 1931 to 1960. The water entering the area *via* precipitation (825 mm per year = 100%) comes from evaporation from the ocean and evapotranspiration from the land. *1* evaporation of intercepted water (precipitation retained by the crowns of trees); *2* uptake of intercepted water (directly, through the wet surfaces of leaves and shoots); *3* "through-fall" through the crowns of trees; *4* stem flow

contribution (75% of the evapotranspiration component, and 45% of the combined sources of precipitation). Superficial drainage and the gravitational water reaching the groundwater level return to the sea what has been carried inland as water vapor. In this return trip, some of both the surface and groundwater may enter the cycle of water use by man; according to recent calculations this represents about 7% of the annual precipitation (16% of the water returning to the sea). For such a highly industrialized area this fraction appears rather modest. But the significant effect of human water consumption—in industry, agriculture, and the household—upon the water cycle lies not so much in the amount involved as in the pollutants added to the waste water, which eventually mix with the rest of the drainage water and in due course produce biologically undesirable effects.

The Earth's Water Balance and Its Significance for Vegetation

Distribution of the Water Supply

The chief reservoirs of water on earth are the oceans: They contain more than 97% of the planet's water (around $1.4 \cdot 10^{18}$ t H_2O). About 2% of the water is stored in frozen form, as ice and snow on the polar caps and in glaciers. The water on the continents (a little over 0.6%) is, for the most part, groundwater; only 1% of it lies so close to the surface that it can be reached by the plant roots, while all the rest sinks to depths of hundreds of meters. The water suspended above the continents and oceans of the world in the form of clouds, fog and water vapor is a negligibly small fraction of the earth's total store—no more than 0.001%.

Water Circulation and the Global Water Balance

The amounts of water stored in various forms are in a complex state of equilibrium with one another. From the surface of the ocean more water evaporates than is precipitated, and the excess water vapor is carried to the land. On a world-wide average, the land areas lose by evaporation less water per unit area than they receive—in some places a great deal less—because their surfaces are not equivalent to an open-water surface. Moreover, the total area of the continents is much less than that of the oceans. Balance-sheets set up by several geophysicists and meteorologists for the earth's water indicate a contribution from ocean to land amounting to around 40,000 km^3 ($4 \cdot 10^{13}$ t) of water per year; the same amount returns each year *via* the rivers. This computation also shows that on a world-wide scale the moisture coming from the ocean makes up only about 40% of the precipitation falling on the continents. All the remainder is provided by evapotranspiration from the continental surfaces, particularly the plant cover; evaporation from the soil surface averages 5—20% of the total evapotranspiration. Because of the high rate of turnover (water vapor remains in the atmosphere for 10 days, on the average), water circulation is quantitatively the most significant cycle of matter on the earth. It also involves the most important energy turnover on earth, since the greater part of the energy in the sun's radiation absorbed by the earth's surface is used in evaporating water. This part of the cycle is closed as precipitation returns the atmospheric water to the surface.

Humid and Arid Regions

On both the oceans and the land there are regions with higher or lower rates of evaporation, and areas with a surplus or a shortage of precipitation. Strong insolation promotes evaporation; in the tropics the amount of water given off annually by the oceans, under the influence of the subtropical high-pressure regions, reaches 2,000 mm or more. On the continents the rate of evaporation is greatest in the equatorial zone, where precipitation is abundant; in the water-poor subtropics and in the higher, cold latitudes there is less evaporation.

Evaporation from land areas must be considered primarily in relation to precipitation. The crucial factor in the water balance of a region is not so much the absolute amounts of precipitation and evaporation but rather the relationship between the two. If annual precipitation exceeds annual evaporation, the region is considered *humid*, and in the reverse case it is *arid*. About $^1/_3$ of the continental area of the earth has a rain deficit, and half of this (about 12% of the total land area) is extremely arid—i. e., the annual precipitation is below 250 mm and the evaporative power of the air exceeds 1,000 mm per year. On the other hand, "perhumid" areas, with a considerable excess of precipitation, account for less than 3% of the continental area (Fig. 112). Extensive dry regions are found mainly between 15° and 30° north and

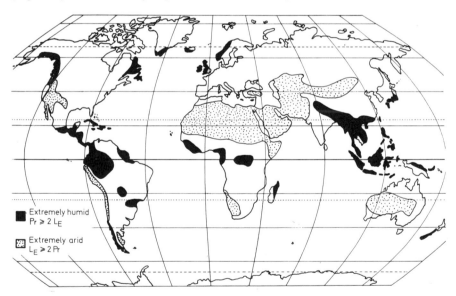

Fig. 112. The distribution of extremely humid and extremely arid regions on the earth. *Extremely humid:* annual precipitation at least twice the amount of water evaporated annually. *Extremely arid:* annual evaporation at least twice as great as annual precipitation. The demarcation of extremely humid and arid regions is based on the maps of Geiger (1965), giving amount of precipitation and actual evapotranspiration. Compare this illustration with the maps of precipitation to be found in any atlas, and with Fig. 4

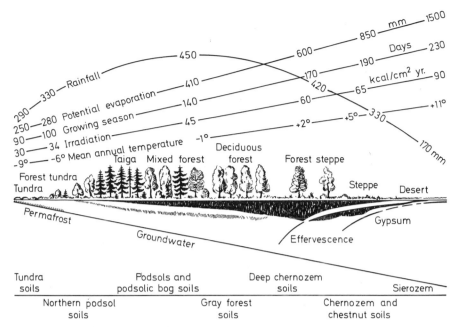

Fig. 113. Diagram of the variations in climate, the distribution of vegetation and the types of soil along a profile through eastern Europe from northwest to southeast as far as the Caspian lowlands. The intersection of the rainfall curve with that for potential evaporation marks the boundary between humid (left) and arid (right) climates. Black: humus horizon; cross-hatched: illuvial B horizon. (From Schennikow as modified by Walter, 1970)

south latitude, and on the lee sides of high mountain chains intercepting rain-bearing winds. In lands far from an ocean there is a gradual transition from a humid climate, through a semiarid intermediate region having occasional or periodic dry periods, to an arid region characterized by permanent drought and increased salinity of the soil (Fig. 113).

The relationship between annual precipitation and annual evaporation gives only a rough indication of the humid or arid character of an extensive area. As far as the plants growing there are concerned, the important thing is that there should be an assured water supply at the time of greatest need—during the growing season. To provide a picture of *relatively humid and relatively arid seasons*, one can construct climatic diagrams like those suggested by H. Walter (Fig. 114). In these, the monthly totals for precipitation and the monthly average air temperature are plotted for the year, the scale being such that 1° C corresponds to 2 mm precipitation. In such a plot, the temperature curve serves as an indicator of the progressive change in evaporative power of the air during the year. That part of the year during which the precipitation curve lies below the temperature curve is a time of drought for the majority of plants that are not irrigated and cannot utilize the groundwater.

Limited precipitation does not in itself bring about aridity; the cold polar regions have little precipitation but are not arid, since there the evaporative power of the air is low. Conversely, precipitation in solid form cannot be taken up immediately by

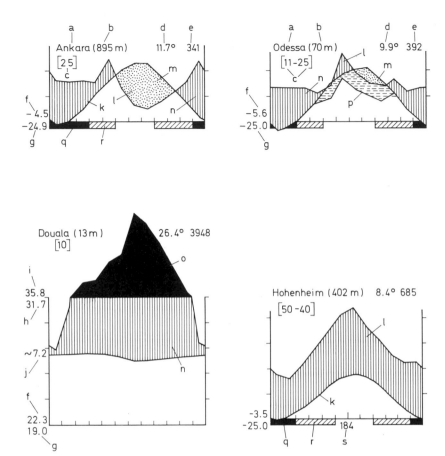

Fig. 114. Climatic diagrams for Stuttgart-Hohenheim (temperate zone, humid), Odessa (temperate zone with summer dryness), Ankara (warm-temperate zone, Mediterranean type of climate with winter rains and summer drought) and Douala (rainy tropics). *Abscissas:* months. *Ordinates:* one division represents 10° C or 20 mm rain. *a* station, *b* height above sea level, *c* duration of observations in years (of two numbers, the first indicates temperature, the second precipitation), *d* mean annual temperature in °C, *e* mean annual precipitation in mm, *f* mean daily minimum in the coldest month, *g* lowest temperature recorded, *h* mean daily maximum in the warmest month, *i* highest temperature recorded, *j* mean daily temperature fluctuation, *k* curve of mean monthly temperature, *l* curve of mean monthly precipitation, *m* relatively dry season (dots), *n* = relatively humid season (vertical shading), *o* mean monthly precipitation >100 mm (black, scale reduced by a factor of 10), *p* precipitation curve with scale reduced (30 mm = 10° C) and dry period indicated by dashed area above, *q* cold season (i.e., months with mean daily minimum below 0° C; black), *r* months with absolute minimum below 0° C and the occurrence of late or early frosts (diagonal shading), *s* mean duration of the frost-free period in days. (From Walter, 1970)

the plants. Winter, for plants projecting above the *snow cover*, is thus not only a cold but also a dry season. Under a covering of snow, on the other hand, there is the equivalent of a maritime environment; snow-covered plants are protected both from evaporation and from extreme temperature fluctuations (cf. p. 194). Where the snow remains on the ground for long periods, however, the growing season is shortened and there are other disadvantages—e.g. infestation by fungi. In the northern hemisphere north of 60° latitude it snows more often than it rains; in the eastern parts of the continents of the northern hemisphere, where the masses of cold air reach further south, this boundary line is shifted nearly to 50° north latitude. In mountains, too, at increasing altitudes an ever greater fraction of the precipitation is in the form of snow.

Effects of Temperature

Plant Temperatures and Energy Balance

Plants are poikilothermic organisms—that is, their own temperatures tend to approach the temperature of their surroundings. However, this matching of temperatures is not precise. It is true that the heat liberated in respiration (about $2 \cdot 10^{-3}$ cal $\cdot$ cm$^{-2} \cdot$ min^{-1}) and that involved in syntheses plays hardly any ecological role, but the temperature of the parts of plants above ground can deviate considerably from the air temperature as a result of energy exchange with the environment (Fig. 115). The heat exchange of plants must therefore be considered with respect to the energy budget in the habitat.

Insolation—The Source of Thermal Energy

Energy flows to earth in the form of radiation from the sun (insolation) which is absorbed by the plant cover, the soil, and the surfaces of the water. Opposing this gain in energy is a loss associated with the thermal radiation from terrestrial bodies. Thermal radiation, in correspondence to the relatively low surface temperature of the earth, is in the long-wavelength range (3—100 μm) and is strongly absorbed by dipole molecules in the atmosphere, particularly water vapor. As a result the air enveloping the earth is warmed, and most of the captured radiation is returned to earth as long-wavelength reradiation from the atmosphere. The net radiation balance

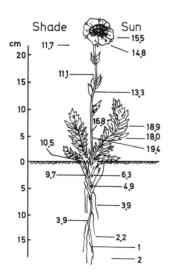

Fig. 115. Temperature distribution in a plant of the arctic species *Novosieversia glacialis* on a sunny July morning when the air temperature was 11.7° C. (After Tichomirow as cited by Walter, 1968)

Q_I at a surface at a given time can be computed from the short-wavelength (0.3—3 μm) balance $\bar{I}_s$ and the long-wavelength thermal radiation balance $\bar{I}_l$.

$$\left(\text{NET RAD. BAL.} \right) \leftarrow Q_I = \bar{I}_s + \bar{I}_l \tag{47}$$

$$\bar{I}_s = I_d + I_i - I_r \tag{48}$$

$$\bar{I}_l = I_a - I_g \tag{49}$$

where I_d = direct solar radiation (cf. p. 000)
I_i = diffuse skylight and cloudlight
I_r = reflected short-wave radiation
I_a = reradiation at thermal wavelengths from the atmosphere
I_g = long-wavelength thermal radiation from ground and plants

The short-wavelength radiation balance received by the earth represents a gain in energy; the thermal radiation balance, as a rule, is a loss, so that the overall radiation balance is positive as long as the short-wavelength contribution predominates. This can apply of course only during the part of the day beginning shortly after sunrise and ending shortly before sunset. The balance becomes negative as soon as the daylight input is insufficient to compensate for the thermal radiant loss. The surplus energy is used in the biosphere for photosynthesis by plants, in warming the phytomass, soil and air, and in phenomena associated with evaporation.

The Thermal Balance of the Plant Cover

Net radiation, energy consumption, and heat exchange are the main factors affecting the energy and hence the thermal balance of plants. The relationships are summarized in the *energy budget equation*:

$$Q_I + Q_M + Q_P + Q_{Soil} + Q_H + Q_E = 0 \tag{50}$$

The terms in this equation, which must counterbalance one another in the overall energy budget of plants, are as follows:

1. *The Radiation Balance Q_I*

2. *The Energy Turnover in Metabolic Processes Q_M*
Under insolation the capture of energy by photosynthesis predominates, while in the dark and in tissues without chlorophyll energy is liberated in respiration. Despite the eminent significance of basal metabolism to the organisms, the share of Q_M in the overall energy turnover is negligibly small, of the order of 1—2%.

3. *Heat Storage by the Phytomass Q_P*
Captured energy can be temporarily stored in the plants—for example, when more energy is taken up by the vegetation stratum than is given off to the atmosphere or to the soil. As a result, the temperature of the plants rises in accordance with the heat capacity of their mass.

4. Heat Storage in the Soil Q_{Soil}

In places free of vegetation, and where the plant growth is sparse (deserts, dunes, mountains, ruderal habitats, clear-cut woods and newly plowed fields), a considerable part of the energy absorbed under insolation is conducted into deeper layers of the soil. Depending on the color, the content of water and air, the structure and the composition of the soil—and even more upon the slope and exposure of the surface—the ground is warmed to different degrees. Compact, wet soils conduct and store heat better than dry soils, so that the latter become hotter on the surface. Extremely high surface temperatures are produced in mountains at sites where the sun's rays fall nearly perpendicularly onto exposed areas of dry dark raw-humus soils; there maximal temperatures exceeding 70° C have been measured. During the diurnal phase of net energy loss from the ground (that is, at night) the direction of heat transport in the soil reverses; the heat stored over the daylight hours is conducted back to the surface of the ground, which during the night has become steadily cooler. There are thus diurnal temperature fluctuations in the upper soil, down to depths of about 50 cm. In regions with a seasonal climate there is a superimposed annual fluctuation of temperature, which can be demonstrated even at depths of several meters (Fig. 116). Under a closed cover of vegetation, the soil is shielded from strong irradiation and radiant-energy loss. Even in the upper layers the diurnal temperature fluctuation is relatively small, and at depths below 30 cm it becomes insignificant (Fig. 117). Moreover, a thick covering of snow can have a similar effect. On the whole, the soil acts as a thermal buffer in the heat balance of a habitat, taking in considerable amounts of heat during the day only to give them up again at night.

5. Exchange of Energy with the Surroundings Q_H and Q_E

Exchange of heat between plants and their environment is also effected by *heat conduction and convection* (sensible heat exchange Q_H) and by *evaporation or condensation* (latent heat exchange Q_E). In neither the plant nor the community is conduction a significant factor; the redistribution of heat by convection and evaporation is mediated chiefly by mass flow of air and water.

Convection

Under conditions of positive radiation balance, the direction of heat convection is usually away from the surface of the plants (Q_H is negative); on the other hand, if the plant surface is cooler than the air, heat is transferred to it from the environment (Q_H positive). Heat exchange with the surrounding air by convection is the more effective, the smaller and more subdivided the leaves and the higher the wind velocity (Fig. 118). Under strong insolation the plant is enclosed in a superheated envelope of air next to its surface. In quiet air a boundary layer centimeters thick can form, from which eddies of warm air rise because they are lighter than the cooler layers of air at a distance from the leaf. The air over an insolated surface is therefore in a state of constant turbulence, and there can be a brisk exchange of air (Fig. 119). Wind sweeps away the boundary layer to within a few millimeters of the plant surface, thus increasing the rate of heat exchange.

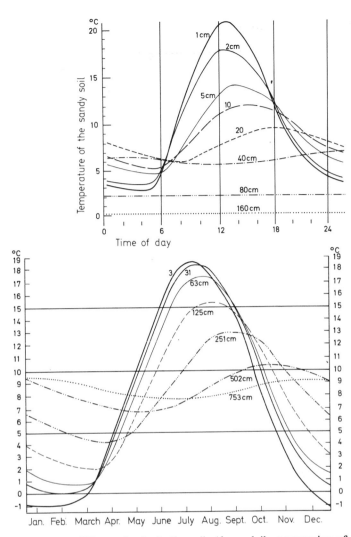

Fig. 116. Variations in temperature at different depths in the soil. *Above:* daily progression of temperature in May in a sandy soil near Pavlovsk. *Below:* annual progression of soil temperature, averaged over 13 years, in Königsberg. (After Schmidt and Leyst as cited by Walter, 1960)

Evaporation and Thermal Balance

The thermal balance of plants is not determined by physical factors alone; because of the stomatal regulation of transpiration, it is also affected by physiological processes in the plant.

Q_E is negative when the plants transpire, and positive when dew or hoarfrost condenses on the leaves. The cooling effect of evaporation can be calculated from the thermal balance equation or by reference to the rate of transpiration. As water is

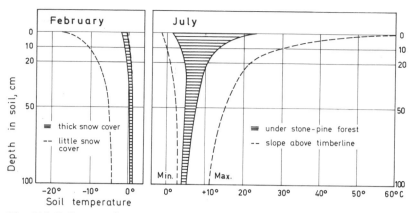

Fig. 117. Influence of snow cover and forestation upon the range of temperature fluctuation (curves show absolute maxima and minima) during the day at different depths in the soil; the habitats are in the region of the alpine timberline and above it. (From Aulitzky, 1961, 1962)

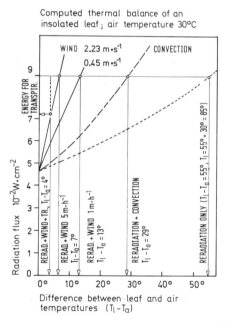

Fig. 118. Diagram of the energy turnover and the heat exchange of a leaf 1 cm in length under irradiation by 1.3 cal · cm^{-2} · min^{-1} (9.1 · 10^{-2} W · cm^{-2}) at an air temperature of 30° C. The leaf temperature is determined by the energy uptake by absorption of radiation and the energy loss involved in reradiation, heat transfer by convection, air movement (two wind velocities shown), and transpiration (heat of vaporization of water). Here transpiration is considered to use energy at the rate of 0.23 cal · cm^{-2} · min^{-1} (corresponding to an average rate of transpiration of 40 mg · cm^{-2} · min^{-1}). The amount by which the temperature of the leaf exceeds that of the air is given by the intersection of the curve for each case with a horizontal line set at the energy level of the incident radiation. With irradiation amounting to about 0.65 cal · cm^{-2} · min^{-1} the leaf, at an air temperature of 30° C, would not be warmer than the air. (After Gates, 1965)

given off, the amount of energy lost is that required to vaporize it. Thus one finds in the literature, as an alternative to Q_E, the expression $\lambda \cdot E$, where E designates the amount of water evaporated and λ is the heat of vaporization of water ("latent heat"). Q_E is the product of the rate of evaporation and the heat of vaporization of water (see p. XIII). Transpiration at a rate of 1 g H_2O · dm^{-2} · h^{-1} (0.1 mm H_2O · h^{-1}) corresponds to energy loss from the plant of 0.1 cal · cm^{-2} · min^{-1}. Cooling by

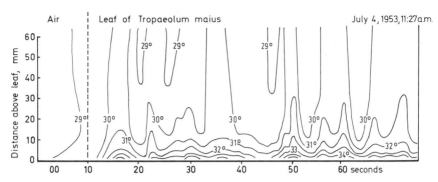

Fig. 119. Isotherms indicating temperature distribution in the air above a horizontal, insolated, transpiring leaf of the common "nasturtium". First, 5 thermocouples shielded from radiation were mounted one above another at 1-cm separations and the temperature distribution in the open air was measured (29° C at time 00). 10 sec later the leaf was pushed under the measurement apparatus and the short-term changes in the temperature profile of the layer of air near the leaf were recorded. One can see that a thin superheated layer forms just at the surface of the leaf, swirls upward after a few seconds, and is reformed from below. (After Berger-Landefeldt, 1958)

evaporation is particularly effective when the air temperature is high, the humidity is low, and the plants are well supplied with water. From Table 36 it can be seen that under steppe conditions, one quarter of the heat lost by a leaf is accounted for by transpiration. If transpiration is accelerated by wind, it extracts so much heat that the leaves can become several degrees cooler than the air. The influence of transpiration upon thermal balance can be readily demonstrated by preventing the plant from giving off water. If one cuts through the petiole of a leaf in order to interrupt the supply of water, or treats the plant with antitranspirants (substances that inhibit transpiration), the leaf temperature rises within a few minutes (Fig. 120).

Plant Temperature and Ambient Temperature

Under strong insolation, the removal of heat by convection and evaporation is frequently insufficient to match the shoot temperature to that of the air. Then heat is stored and the leaves warm up by as much as 10° C—in exceptional cases even 15 or 20° C—above the temperature of their surroundings. Massive plant organs with poor heat exchange, such as succulent leaves and stems, fruits, or tree trunks can even on occasion reach temperatures of more than 20° C above air temperature. The heating of plant parts varies according to their orientation with respect to the incident radiation and even more greatly as a function of their exposure to wind, so that (for example) not only do the individual leaves have different temperatures (cf. Fig. 115) but even within a leaf the temperatures may vary (Fig. 121). The same holds for the cooling of plants at night. When there is no wind, the temperature at the margins and tips of the leaves is lower than that of the blades. Dew and hoarfrost are deposited first on the cooler parts, and these parts freeze sooner than the remainder of the leaf (in cases of "radiation frost").

Table 36. Difference between the temperature of a *Canna* leaf and the air temperature, and the cooling effect of transpiration. (From Raschke, 1956)

| | Heat exchange by convection and wind | | | |
| | very low | | very high | |
	$T_{leaf}-T_{air}$	cooling by transpiration	$T_{leaf}-T_{air}$	cooling by transpiration
Day (Radiation balance: $0.7 \cdot 10^{-2}$ W $\cdot$ cm^{-2})				
Tropical day 35°C, 90% *RH*	+26.4°C	*−3,0°C*	+1.9°C	*−0.4°C*
Steppe day 40°C, 7% *RH*	+20.9°C	*−4.8°C*	−1.1°C	*−3.0°C*
Temperate day 22°C, 61% *RH*	+15.9°C	*−1.3°C*	+0.7°C	*−0.5°C*
Night				
Tropical night (Radiation balance: $-0.35 \cdot 10^{-2}$ W $\cdot$ cm^{-2}) 30°C, 94% *RH* Dew point: 29.0°C	− 1.5°C	*+0.9°C*[a]	−0.2°C	*−0.1°C*
Steppe night (Radiation balance: $-0.56 \cdot 10^{-2}$ W $\cdot$ cm^{-2}) 20°C, 23% *RH* Dew point: −1.9°C	− 4.3°C	*−0.1°C*	−0.5°C	*−0.2°C*
Temperate night (Radiation balance: $-0.49 \cdot 10^{-2}$ W $\cdot$ cm^{-2}) 14°C, 95% *RH* Dew point: 13.2°C	− 2.4°C	*+1.3°C*[a]	−0.3°C	*0*

[a] Heat transfer to leaf by condensation

Example of the interpretation of the table: Under the conditions prevailing during a tropical day the leaf warms up by 26.4°C in still air, while with adequate heat exchange its temperature is only 1.9°C above that of the air; if it were not cooled by transpiration, the temperature of the leaf would be 3°C higher than that of the air in the first case, and 0.4°C higher in the second.

In the crowns of trees and in tall stands of plants, energy exchange occurs chiefly in a narrow zone near the upper surfaces (Fig. 122). In this active layer (the "*effective surface*" of the stand), the thermal individuality of the different plant parts is most clearly evident and the temporal variation of temperature is greatest. In this region the leaves and twigs warm up the most during the day and cool off most rapidly after sunset. At the surface of the crown the daily fluctuation of temperature is

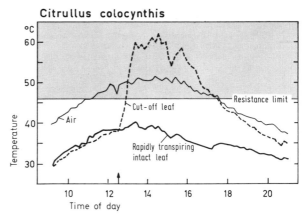

Citrullus colocynthis

Fig. 120. Cooling effect of transpiration upon the leaves of a watered *Citrullus* plant under desert conditions. During rapid transpiration the leaves, despite intense insolation, are much cooler than the air. If a leaf is cut off (arrow) so as to make vigorous transpiration impossible, the leaf temperature rapidly rises above that of the air, becoming so high that signs of heat injury appear (the range of temperatures associated with heat injury is shown in gray). Plants like *Citrullus*, which ordinarily maintain a temperature lower than that of the air, can survive in hot habitats only if they are able to transpire at a high rate. (After Lange, 1959)

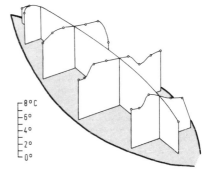

Fig. 121. Distribution of temperature over a leaf of *Canna indica* at noon. Radiation balance = $3.39 \cdot 10^{-2}$ W $\cdot$ cm^{-2}; wind velocity, 3.3 m $\cdot$ s^{-1}; mean temperature of the leaf 5.5.°C above that of the air. (After Raschke, 1956)

greater by 2—4° C, on the average, than in the open air. Under a roof of leaves or in a dense stand of plants, the intensity of radiation declines rapidly and atmospheric reradiation is limited by the shielding leaves. The entire process of energy conversion is damped, so that within the stand more stable temperatures prevail. In dense tropical forests the temperature hardly fluctuates at all.

The Energy Budget of the Biosphere

The rules governing thermal balance, described above with respect to stands of plants, are valid for the entire biosphere as well. The balance of radiation and temperature of the earth is a subject of research in the field of geophysics; the energy-balance equation (Formula 50) is an application to plant ecology of the formalism used to compute the thermal balance of entire regions. For geophysical purposes

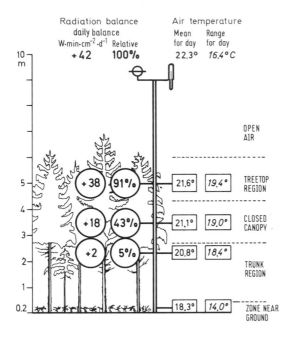

Fig. 122. Radiation balance and air temperature (daily mean and daily fluctuation) over and within a dense young spruce forest on a typical day of fine weather in midsummer. The most pronounced turnover of energy takes place in the upper part of the treetop, where the daily air-temperature fluctuation is also greatest. Near the ground, where only a very small part of the energy is converted, the air temperature remains stable and cool. (Cf. also Fig. 8). (After Baumgartner as cited by Geiger, 1961)

the negligibly small term representing energy turnover by metabolic processes (Q_M) is omitted, whereas the flow of heat in the soil (Q_{Soil}) becomes more significant. The relationships between Q_P, Q_{Soil}, Q_H and Q_E vary in characteristic ways from one place to another, depending upon type and degree of closure of the plant cover (Fig. 123). Over long periods of time (years) the currents of heat in soil and vegetation have average values near zero, so that in the equation representing the net energy budget for a year, only the terms Q_L, Q_H and Q_E remain. The ratio Q_H/Q_E (the Bowen ratio) depends upon the density of plant growth and the availability of water. In the vicinity of extensive water surfaces and of stands of plants transpiring at a high rate, a large part of the radiant energy is accounted for by evaporation during the day; the lower layers of air are warmed to only a moderate degree. The maintenance of thermal and water balance in a closed plant cover thus tends to make the local climate milder, and creates more favorable conditions for life in the biosphere, than would be the case if the plants were scattered or absent.

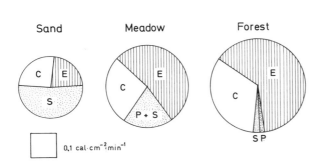

Fig. 123. Energy turnover in a field of sand, a meadow and a dense spruce forest with the sun high in the sky. C convection; E evaporation; S heat flow in the soil; P heat storage in the plant cover. (After Albrecht and from Baumgartner as cited by Berger-Landefeldt, 1967)

198

The Effects of Temperature upon the Vital Processes of Plants

Life-Supporting Range and Functional Range

Sufficient but not excessive heat is a basic prerequisite for life. Each vital process is restricted to a certain temperature range and has an optimal operating temperature, on either side of which performance declines. Thus for each plant species, and for each stage of development, one can determine characteristic "cardinal temperatures". These are not rigid constants but rather ranges about the genetically fixed norm within which the optimal, minimal and maximal temperatures for the plant can shift as the plant adapts to environmental conditions. Terrestrial vascular plants, as a rule, can thrive within a wide range of temperatures and are called *eurythermic*. For such plants, in an active state, the life-supporting temperature range usually extends from −5° C to about +55° C—a span of 60 degrees; such plants are productive, however, only between about 5° C and 40° C (Fig. 124). Among the aquatic plants, particularly the thallophytes, there are *stenothermic* species, specialized for life in very narrow—but sometimes extreme—temperature ranges. For example, snow and ice algae (e.g., *Chlamydomonas nivalis* on snow fields in the high mountains, and various green algae and diatoms in the polar ice) can thrive only near the

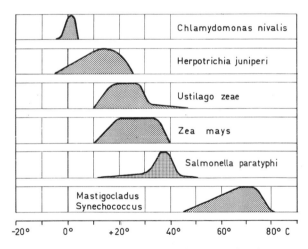

Fig. 124. The temperature ranges in which various plants thrive. Plants restricted to low temperatures are *Chlamydomonas nivalis* and other snow algae, as well as the mold *Herpotrichia juniperi*, which infests the snow-covered twigs of conifers. The mycelial growth of the corn smut *Ustilago zeae* is largely determined by the temperature range for activity of the host plant. Parasites of warm-blooded animals, like *Salmonella paratyphi*, display an especially narrow optimum for development. At extremely high temperatures the only organisms that can thrive are thermophilous bacteria and blue-green algae (for example, *Mastigocladus* and *Synechococcus* species from geysers). (From data in Went, 1957; Altman and Dittmer, 1966; Raeuber *et al.*, 1968; Kol, 1968; Brock, 1967; De Wit *et al.*, 1970; Müller and Löffler, 1971)

199

freezing point. Stenotherms also include many parasitic bacteria and fungi that are adapted to the temperatures at which infection and spread through the host proceed optimally.

For ecological purposes it is necessary to know, for the different plant groups, the following thermal requirements and limits of tolerance:

1. The Survival Limits

These are the lowest and highest temperatures at which a plant can survive. A distinction is made between the activity limit and the lethal limit. When the *activity limit* is exceeded the active vital processes are reversibly slowed to a minimal rate and the protoplasm enters a state of anabiosis (under excessive heat or cold). At the *lethal limit* permanent injury is sustained and life is extinguished. Resistance to great heat or cold is an advantage to any plant, but especially to those that must avoid competition; these cannot establish themselves under favorable temperature conditions and are found only in open, and therefore climatically extreme, habitats.

2. The Temperature Range for Dry Matter Production and Growth

Basal metabolism and the synthesis of new tissue are the prerequisite for growth and development and thus are crucial determinants of the competitive ability of a species. Examples of the influence of temperature upon nutrition and metabolism have been given previously. Shoot growth in most eurythermic plants proceeds optimally between 20° and 25° C; moreover, depending on the range of distribution of the species, an alternation between different daytime and nighttime temperatures may be advantageous. The latter phenomenon reflects an adaptation to the daily temperature rhythm (thermoperiodism; cf. p. 219). Plants of continental regions where temperature oscillates widely in the course of a day develop best if the night is about 10—15° C cooler than the day, whereas maritime species prefer a difference of 5—10° C (Fig. 125). Some tropical crop plants such as sugar cane and peanut thrive in the absence of such oscillations, and there are certain species (e.g., the African violet, *Saintpaulia*) that actually prefer a higher nighttime temperature. Stem growth in general is affected more by the nighttime temperature than by the temperature during the daylight hours.

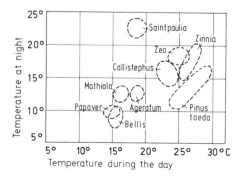

Fig. 125. Optimal temperature range for growth and development in various plants. As a rule, plants grow best if the daytime temperatures are higher by a few degrees than those at night; an exception is the African violet (*Saintpaulia ionantha*), which does best under higher nighttime temperatures. (After Went, 1957, with inclusion of data from Kramer, 1958)

200

3. The Temperature Limits and Heat Requirements for Reproductive Processes

If a population is to maintain itself, and a species to expand its range, it is not sufficient that the individual plants simply withstand extreme situations and perform their vegetative functions adequately; the requirements for flowering and the ripening and germination of seeds must be met. The range of temperatures suitable for the latter processes differs in many cases from that for growth and the development of the vegetative organs.

The cardinal temperatures for germination of spores and seeds must correspond to external conditions guaranteeing sufficiently rapid development of the young plants. The temperature range for *initiation of germination* is wide in species that are broadly distributed and in those adapted to wide temperature fluctuations within their habitat. Tropical plants germinate optimally (i.e., the percentage of seeds that germinate is greatest) between 15° and 30° C; the range for plants of the temperate zone is 8—25° C, and for alpine plants, 5—30° C. The *rate of germination* rises with increasing temperature. In species germinating in summer (usually those of northern origin), as opposed to winter-germinating species (from regions with mild winters), germination proceeds extremely slowly at low temperatures; only after the seed-bed has warmed up to more than 10° C does the process accelerate, but then it soon makes up for lost time (Fig. 126). Thus synchronization is achieved with the season most favorable for development of the young plants. In some plants there are complicated mechanisms to prevent germination at unfavorable times. The seeds of many woody plants in cold-winter regions (a number of forest trees and Rosaceae) and the seeds of mountain plants (e.g., *Silene acaulis* and many Primulaceae) germinate more readily if they have been exposed, while in a turgid state, for a considerable time to low temperatures or frost.

The *initiation of flowering* is induced only within a narrow temperature span, and still different temperatures are effective in bringing about the development and unfolding of the flowers. Winter annuals and biennials, as well as the buds of certain woody plants (e.g., the peach) require a cold winter season in order to flower normally in the spring ("chilling requirement"). They do not become ready to flower until they have been exposed for weeks to temperatures between —3° and +13° C,

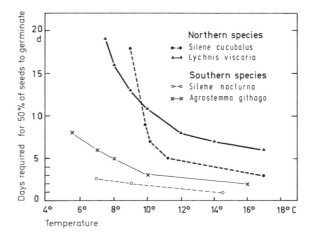

Fig. 126. Temperature dependence of rate of germination in various species of Caryophyllaceae. The northern species germinate in summer, and germination is greatly inhibited below 10° C; the southern species germinate in winter, and the process is rapid even at low temperatures. (After Thompson, 1968)

ideally between +3° and +5° C. This acquisition or enhancement of the ability to flower by exposure to cold is called *vernalization*. If the cooling period is too short, comes at the wrong time, or is interrupted by warming above 15° C, the effect does not appear.

Fruits and seeds require more heat to ripen than is necessary for the completion of growth by shoots and roots. In habitats with both shorter and cooler growing seasons, a plant species can better maintain itself if it has alternative means of propagation, such as the formation of runners, spread by rhizomes, or other vegetative mechanisms. The coldest regions, like the dryest, are colonized almost exclusively by cryptogams, in which the elaboration of reproductive organs is minimal.

The Temperature Limits for Plant Life

Extremes of Temperature on the Earth

Only under water and deep in the ground is the temperature restricted to the range between about 0° C and 20—25° C, where plants are in no danger and their vital functions are promoted. Near the soil surface on the continents, as well as in the tidal zones and in shallow waters, the temperature oscillates with the time of day (and, outside the equatorial zone, with the seasons) between limits that can be potentially lethal.

Particularly *high* air temperatures occur at latitudes near the northern and southern tropics; as an extreme case, maxima of +57° to +58° have been measured in Libya, Mexico and California. Over about 23% of the continental area of the earth, mean

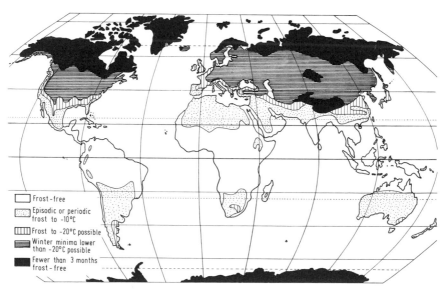

Fig. 127. The occurrence of sub-freezing temperatures on the earth. (From data of Walter and Lieth, 1967; Blüthgen, 1964)

annual air temperature maxima of over 40° C are to be expected; in such areas plant temperatures of 50° C or more can occur in intense sunlight. Apart from these hot regions, severe local overheating to 60° and 70° C is possible, primarily on rocks and on open habitats sloping toward the sun. The hottest spots on earth inhabited by living organisms are geysers, in which the water coming to the surface is at 92–95° C; near these, some bacteria colonize zones as hot as 90° C.

The *coldest* temperatures measured on the earth have been in Antarctica (near −90° C); in Greenland and in eastern Siberia air-temperature minima between −66° and −68° C have been observed. Relatively severe frost (a mean annual air temperature minimum below −20° C) is to be expected over 42% of the earth's surface, and only a third of the land area never experiences freezing temperatures (Fig. 127). An important factor in the effects of periods of frost is whether they occur regularly in a region as the seasons alternate, or whether they are isolated episodes. Episodic frosts are usually more dangerous even though the temperatures are rarely very low, since they may catch the plants in a sensitive phase. In contrast, plants can "prepare" for the regular annual return of winter frosts, by allowing for a gradual hardening of their vegetative processes to low temperatures so that they suffer no damage.

Effect of Temperature Stress

Heat and frost impair the vital functions and limit the distribution of a species, depending upon their intensity, duration and variability; the state of activity and degree of hardening of the plants are even more important factors. Stress is the *exposure to extraordinarily unfavorable conditions*; they need not necessarily represent a threat to life, but they do trigger an "alarm" response in the organism if it is not in a dormant state. Resting stages such as dry spores, and poikilohydric plants in a dry state, are insensitive and can survive undamaged any temperatures occurring naturally on the earth (Table 37).

Protoplasm responds to stress with an initial feverish acceleration of metabolism. The increased respiration observable as a stress reaction (cf. Fig. 36) is an expression of the effort being made to repair damage incurred and to make the adjustments in fine structure necessary for adaption to the new situation. The stress reaction amounts to a race between the adaptive mechanisms and the destructive processes in the protoplasm that would lead to death.

Cell Death by Heat and Cold

When critical temperature thresholds are exceeded, cell structures and functions may be damaged so abruptly that the protoplasm dies immediately. In nature this sort of destruction often occurs during episodic frosts—for example, late spring frosts. But damage can also come about gradually, as the equilibria of certain vital processes are affected and their operation is impaired; finally some functions essential for life cease and the cell dies.

Symptoms of Temperature Injury

Temperature-sensitivity varies with the vital process concerned (Fig. 128). The first effect to appear is the cessation of protoplasmic streaming, since this is directly dependent upon energy provided by respiration and upon the availability of high-energy phosphates. Next, the rates of photosynthesis and respiration are decreased. Photosynthesis is particularly vulnerable to heat, and respiration is disturbed primarily by cold. Plants injured by extreme temperatures, after they are returned to more moderate conditions, respire at strongly fluctuating (and usually abnormally high) rates. Damage to the chloroplasts is followed by a residual, sometimes permanent, inhibition of photosynthesis. In the terminal stage the semipermeability of the biological membranes is lost, so that the cell compartments and particularly the thylakoids of the plastids collapse and the cell sap emerges into the intercellular spaces.

Causes of Death by Heat

Heat causes death by damaging membranes, and in particular by inactivating and denaturing proteins. Even if only some especially thermolabile enzymes are put out of action, so that nucleic-acid and protein metabolism becomes disorganized, the cells eventually die. Soluble nitrogen compounds then accumulate in such high concentration that they seep out of the cell and are lost; moreover, toxic decomposition products are formed and can no longer be made innocuous through metabolic processes.

Death by Cold and Freezing

In cases of cold damage, there is a distinction between thermal injury to protoplasm as such and injury by the process of freezing. Some plants of tropical origin suffer *chilling injury* even when the temperature is several degrees above freezing. Like death by heat, death by cold is the consequence primarily of disturbed nucleic acid and protein metabolism, though changes in permeability and stoppage of the flow of photosynthates are also involved.

Plants resistant to chilling above the freezing point are damaged by frost—that is, by *ice formation in the tissues.* Protoplasts with a high water content that have not been

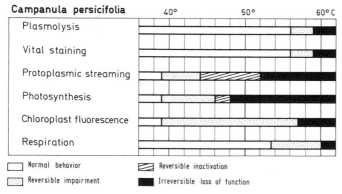

Fig. 128. Impairment of various cellular functions in the leaves of a bellflower as a result of heating for 5 minutes. Protoplasmic streaming and photosynthesis are particularly heat-sensitive functions; respiration and semipermeability are relatively resistant to heat. (After Alexandrov, 1964)

hardened to cold can freeze intracellularly; ice crystals form very rapidly inside the cell and it perishes. Usually, though, ice is formed not in the protoplasts but rather in the intercellular spaces and in the cell walls. This sort of ice formation is called extracellular. As ice crystallizes out it has an effect like that of dry air, since the vapor pressure of ice is lower than that of a supercooled solution. Thus water is withdrawn from the protoplasts (Fig. 129), they shrink markedly (to $^2/_3$ their volume), and the concentration of dissolved substances rises correspondingly. The redistribution between free and bound water and the ice phase continues until water potential equilibrium is reached between the ice and the water in the protoplasm. The water potential at which equilibrium is reached is temperature-dependent; at $-5°$ C it is about -60 bar and at $-10°$ C as low as -120 bar. Freezing thus has the same effect upon protoplasm as desiccation. The frost resistance of a cell is greater if the water is firmly bound to protoplasmic structures, ions and colloids. When water is withdrawn from the protoplasm (whether by drought or freezing), enzyme systems associated with the membranes, which are crucially involved in ATP synthesis and phosphorylation, are inactivated (U. Heber, K. Santarius). This inactivation is caused by an excessive, effectively poisonous concentration of salt ions and organic acids in the solution remaining unfrozen. Sugar, sugar derivatives, certain amino acids and proteins, on the other hand, act to protect the membranes and enzymes from these effects. There are some indications that membrane damage can also result from the denaturation of proteins by freezing (J. Levitt).

Temperature Resistance

As in the case of drought, one can define temperature resistance as the net result of the capacity of a plant's protoplasm to survive extreme temperatures ("tolerance", according to J. Levitt) and the effectiveness of its mechanisms for delaying or pre-

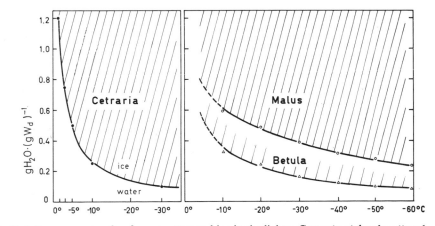

Fig. 129. Relative amounts of unfrozen water and ice in the lichen *Cetraria richardsonii* and in twigs (in winter condition) of the apple variety "Antonovka" and the beech, as the temperature falls. Most of the water freezes above $-10°$ C, but even at lower temperatures water continues to be converted to ice. The extraction of water as a result of freezing affects the protoplasm in the same way as does desiccation. (After Scholander *et al.*, as cited by Levitt, 1956, and from data from Krasavcev, 1968)

venting the onset of damage ("avoidance"). Fig. 130 gives a diagrammatic survey of the components of temperature resistance.

Mechanisms for Avoidance of Injury

There are only a few mechanisms—and these not very effective—by which plants are able to protect protoplasm from extreme temperatures. Insulation from overheating and frost is effective only for short exposures. For example, in the dense crowns of trees and in cushion plants the inner, more deeply located buds, leaves and flowers are less endangered as the air temperature falls below freezing than are the outer parts of the plant; conifer species with particularly thick bark can better withstand fires in the undergrowth of the forest. On the other hand, there are two main mechanisms of general significance: the delay of ice formation in the tissues and the reduction of heating by the reflection of radiation and by cooling through transpiration.

Delayed Ice Formation in the Tissues. Dissolved substances lower the freezing point of solutions. Cell sap, depending on its concentration, freezes between $-1°$ and $-5°$ C; in a tissue the cells first begin to freeze at even lower temperatures than does the cell sap, because not only the concentration-dependent *freezing-point depression* but also other water-binding (e.g., matrix) forces come into play. Moreover, the water in the cells can be supercooled; that is, it can actually be cooled to temperatures somewhat below the freezing point without freezing immediately. This supercooled state, however, is labile. It is seldom maintained for more than a few hours and at most can assist the plant to survive a "radiation frost". These processes of supercooling and freezing can be demonstrated if a temperature decrease in a plant is followed in detail. Tissue temperature at first falls, with only a slight delay as compared with the cooling of the air, down to $-5°$ or $-7°$ C and occasionally even further (to $-15°$ or $-20°$ C). Then there is a sudden rise in tissue temperature as heat of crystallization is liberated; this event indicates the onset of freezing (Fig. 131). The highest temperature that can be measured during ice formation after supercooling is the freezing point of the tissue. Once an equilibrium between liquid and solid

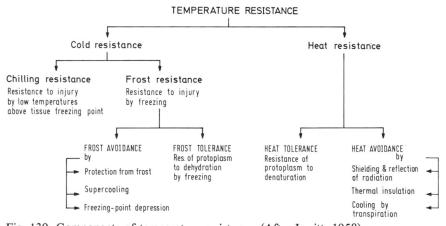

Fig. 130. Components of temperature resistance. (After Levitt, 1958)

phases has been reached, the temperature begins to fall again. This point marks the limit of the range over which protection is offered by these mechanisms.

Freezing-point depression amounts to protection against frost which, though moderate, is reliable; plants depend on it during the growing season. This mechanism is sufficient to enable the evergreen leaves of Mediterrannean and subtropical plants to survive the winter (Table 37). Fully formed leaves are usually safe from freezing down to temperatures of $-3°$ to $-5°$ C, and toward winter the freezing point falls by another $2-5°$ C as osmotically effective cell-sap components accumulate in evergreen leaves. Only young leaves, flowers and juicy fruits freeze in the range $-1°$ to $-2°$ C.

Heat Reduction by the Reflection of Radiation and Cooling by Transpiration. Dangerously high plant temperatures—except of course in fires—occur only under intense insolation. Increased reflection counteracts overheating, as does turning the leaves so that they intercept less sunlight. There are trees among the Caesalpiniaceae which, at air temperatures above $35°$ C, even though they are well supplied with water, fold the blades of their leaves together and thus cut their absorption of radiation in half.

Another protective measure against overheating is cooling by transpiration. As long as water can be supplied to the leaves of desert and steppe plants at a high rate, the leaves remain $4-6°$ C, and in extreme cases as much as $10-15°$ C, cooler than the air; without this heat-reducing mechanism, they would be injured (cf. Fig. 120).

Protoplasmic Tolerance

Under conditions of long and regularly repeated periods of exposure to intense heat and cold, the temperature of a plant must follow that of the environment; then its

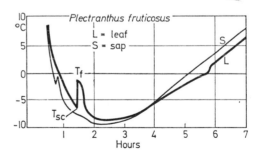

Fig. 131. The change of temperature in a leaf, and in the sap expressed from a leaf, of *Plectranthus* during a freezing experiment. At time 0 the samples were placed in a precooled chamber. There, their temperature rapidly fell to the supercooled level (T_{sc}). The leaf could be supercooled to a lower temperature (about $-6°$ C) than the sap on a piece of filter paper (supercooling limit at about $-2°$ C). As freezing began, the temperature in both leaf and sap suddenly rose as a result of the liberation of heat of crystallization, and the initial subsequent rate of fall was slow. The highest temperature reached in this exothermic process can be taken as the freezing point (tissue freezing point of the leaf, T_f, or freezing point of the sap/filter-paper system). During thawing, the rise of temperature in the leaf is delayed in the range of the freezing point because heat was used in melting the ice. (After Ullrich and Mäde, 1940)

Table 37. Frost resistance (temperature at the first appearance of injury), initial freezing (temperature at the beginning of ice formation) and protoplasmic frost tolerance in evergreen leaves and needles in winter. The frost tolerance corresponds to the difference between the temperature at first appearance of injury and the initial freezing temperature. (From Larcher, 1973)

Plant	Frost injury	Initial freezing	Frost tolerance
Eucalyptus globulus	− 3°C	− 3°C	*none*
Citrus limon	− 5	− 5	*none*
Ceratonia siliqua	− 5	− 5	*none*
Nerium oleander	− 7	− 7	*none*
Olea europaea	−10	−10	*none*
Pinus pinea	−11	− 7	*4°C*
Quercus ilex	−13	− 8	*5*
Cupressus sempervirens	−14	− 5	*9*
Taxus baccata	−20	− 6	*14*
Abies alba	−30	− 7	*23*
Picea abies	−38	− 7	*31*
Pinus cembra	−42	− 7	*35*

survival depends on protoplasmic tolerance. This capacity is genetically determined and thus differs both among species and among varieties of the same species. Moreover, tolerance varies in the course of development of a single plant. Seedlings, the new shoots put out by woody plants during the most intensive growth period in the spring, cultures of microorganisms in their phase of exponential growth, and plants in flower are all extremely sensitive to high and low temperatures.

Frost Tolerance and Hardening to Frost. In regions with a seasonal climate, perennial plants become hardened to frost—that is, they acquire an ability to survive ice-formation in their tissues. This ability is minimal during the growing season, develops gradually during the fall to a high point in winter, and in the spring is lost again (Fig. 132). Only after growth is completed does the plant enter a state of readiness for hardening; the process then advances in phases, with each stage preparing the way for the next. In the theory of frost-hardening developed by I. I. Tumanov, the process is induced in winter grain and fruit trees (these plants have been the most thoroughly investigated) by exposure for several days or weeks to temperatures just above zero. In this *pre-hardening* stage, sugar and other protective substances are accumulated in the protoplasm; the amount of water in the cells falls and the central vacuole divides into a number of smaller vacuoles. At this point the protoplasm is prepared for the next phase, which takes place when the temperature falls regularly to between −3° and −5° C. Now the fine structures and the enzymes of the protoplasm are reorganized in such a way that the cells can withstand the removal of water by ice formation. Only then can the plants enter without danger the *terminal stage of hardening*, so that during prolonged freezing, with temperatures no higher than −10° to −15° C, the protoplasm achieves maximal frost tolerance. The critical temperature ranges vary from one species to another. Birch seedlings in readiness

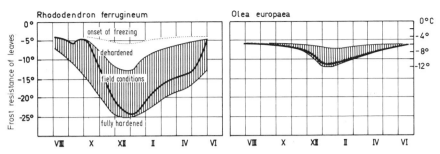

Fig. 132. Annual trends of frost resistance and the onset of freezing in leaves of *Rhododen-dron* and the olive tree (months are indicated by Roman numerals). Characteristics of the individual resistance of a species are the degree of hardening (and the tendency to deharden) in the season when frosts are possible, and the relationship between the onset of freezing (dotted curve) and initial frost damage. Olive leaves suffer damage even in winter as soon as ice is formed, whereas the *Rhododendron* leaves withstand ice formation in winter (they are frost-hardened). The shaded area indicates the range between *minimal resistance*, when frost-hardening has been lost after several days in a warm room, and *maximal resistance*, achieved by stepwise hardening. Between these limits of the possible thermal influence on frost resistance, lies the *actual resistance* of the plants in the field, which depends upon the preceding weather conditions. (After Pisek and Schiessl, 1947; Larcher, 1954, 1963b, and unpublished data)

for hardening, which before the process was begun would have frozen to death at −15° to −20° C, at the end of the first phase can already endure −35° C; when they are completely hardened they can survive temperatures as low as −195° C. Cold thus forces hardening to proceed. Once the most severe cold spell is over, the protoplasm returns to the first stage of hardening, but tolerance can be returned to its highest level by renewed cold periods—as long as the plants remain in the winter state of development.

In the course of the winter, adaptations that promptly adjust the level of resistance to changes in the weather are superimposed on these seasonal variations. Cold promotes hardening particularly in the early winter, when resistance can be brought to a peak within only a few days. Thawing, on the other hand, toward the end of winter causes the plants rapidly to lose their tolerance; but even in midwinter considerable resistance can be lost after a few days at +10° to +20° C. The degree to which frost tolerance can be influenced by cold and heat—the range for *responsive adaptation of tolerance*—is characteristic of individual plant species (Fig. 132).

As soon as a plant has emerged from winter dormancy, its resistance to cold, as well as its ability to develop such resistance, is quickly lost. In spring there is a close parallel between the plant's activity in putting out new growth and the loss of resistance to cold (Fig. 133).

Heat Tolerance. Heat resistance, too, follows an annual cycle in many plants, though the changes are much less extensive than those of frost tolerance. Seasonal changes in heat tolerance are controlled mainly by developmental processes and bear little relation to the actual temperature in the field (Fig. 134). Thus all plants are

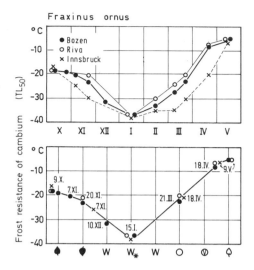

Fig. 133. Annual change in frost resistance of the cambium of twigs of flowering ash, from Innsbruck, Bozen and Riva (on Garda Lake, Italy). In the upper graph the change in resistance is given for each month, and in the lower graph it is related to the phenological state of the twigs. *Symbols* from left to right: leaf coloration, leaf abscission, onset of winter dormancy, main frost period, ending of winter, bud swelling, bud opening, complete leaf expansion. The data points are accompanied by the dates (Roman numeral indicates month) on which they were taken; a given phenological aspect appears in the spring in Innsbruck (beyond the natural range of the species) almost a month later than in Bozen (at the northern limit of the range) and in the more southern habitat near Garda Lake. It is evident that the differences in the curves plotted as a function of time in this species are mainly ascribable to differences in the state of development. (After Larcher and Mair, 1968)

very sensitive to heat during their main growth period. Numerous land plants in the temperate zone acquire their greatest heat tolerance during the winter dormant period. But in addition to this developmentally-determined, ecologically paradoxical phenomenon, there are land plants which also (or only) show an increase of heat tolerance in the summer, and still others with no seasonal change in tolerance at all. The behavior of the algae makes sense from an ecological point of view; both fresh-water and marine algae adapt their heat tolerance to the temperature of the water. In late summer their tolerance is highest, and in winter lowest. The amplitude of this annual oscillation is greater, the greater the annual water-temperature variation.

Plants respond to heat stress with an extraordinarily rapid *responsive adaptation*. Hardening to heat can be achieved within hours, so that resistance on hot days is greater in the afternoon than in the morning. Usually the enhanced resistance disappears rather quickly when it grows cooler, but there are also cases in which the protoplasm remains heat-tolerant, if the plants were raised under continuously high temperatures (heat-acclimatized habitat types).

210

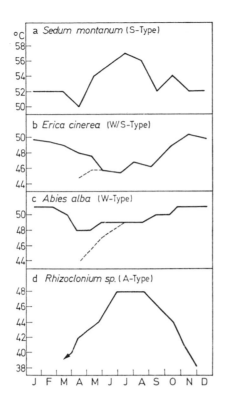

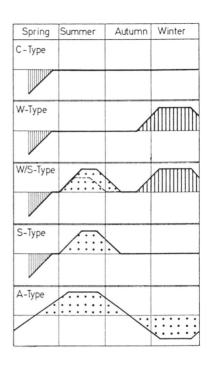

Fig. 134. *Measured seasonal changes* in heat resistance for certain species (left), and a *diagrammatic representation* of the changes found in different types of plant (right). Resistance was measured for 30-minute exposures to heat; the dashed lines refer to young leaves. In the diagrams, resistance changes representing adaptation to the prevailing temperature of the environment are identified by the dotted areas; the reduced resistance during the time of intensive growth is shown by closely-spaced shading; and the increased resistance during winter dormancy, associated with the general enhancement of protoplasmic resistance, is shown by widely spaced shading. Apart from the 5 types shown, there are also plants with no discernable annual change in heat resistance (e.g., *Asplenium ruta muraria* and *Ilex aquifolium*). (From Larcher, 1973)

Constitutional Types of Heat and Cold Resistance

Not all plants can survive low temperatures. Even among frost-tolerant plants not all are capable of passing through all the phases of hardening. Plant species also differ greatly in their sensitivity and ability to develop resistance to heat. Plants can be categorized according to the maximal resistance they can achieve and the roles of avoidance mechanisms and protoplasmic tolerance; such groups are of ecologic significance since specific temperature resistance is one factor limiting distribution (Tables 38 and 39). The measure of resistance is ordinarily taken as the temperature at which half of the plant samples are destroyed (the temperature lethality TL_{50}).

Table 38. Maximal temperature resistance of poikilohydric plants well supplied with water, as compared with the dry, anabiotic state. (From the data of numerous authors; reference lists are given in Christophersen, 1955; Biebl, 1962; Altman and Dittmer, 1966; Larcher, 1973)

Plant group	°C at which cold injury[a] occurs		°C at which heat injury[b] occurs	
	wet	dry	wet	dry
Rickettsias			50—70	
Bacteria				
Plant pathogens			45—56 (60)	
Animal pathogens	—15 to —196		50—70	
Primarily saprophytic bacteria			up to 70	
Thermophilic bacteria			up to 95	
Bacteria spores		to —253	80—120	up to 160
Fungi				
Yeasts	to ca. —20			
Plant pathogens			45—65 (70)	
Primarily saprophytic fungi	0 to <—10		40—60 (80)	75—110
Thermophilic fungi	ca. +5			
Fungus spores			50—60 (100)	over 100
Algae				
Marine algae				
sublittoral (tropics)	+16 to +5		32—35	
sublittoral (cold oceans)	ca. —2		22—26	
tidal zone	—10 to —40 (—70)		36—42	
Fresh-water algae	— 5 to —10 (—20)		40—45 (50)	
Aerial algae	—10 to —30	—196		
Thermophilic algae				
eukaryotic algae			45—50	
Cyanophyceae			70—75	
Lichens	—80 to —196	—196	35—45	70—100
Mosses				
Mosses of the forest floor	—15 to —25		40—50	80—95
Rock mosses	—30	—196		100—110
Ferns	—20	—196	47—50	60—100
Spermatophytes				
Ramonda myconi	— 9	—196	48	56
Myrothamnus flabellifolia		—196		80

[a] After at least 2 hrs exposure to cold.
[b] After 0.5 hr exposure to heat.

Table 39. Temperature resistance of the leaves of vascular plants from different climatic regions. Limiting temperatures are for 50% injury (TL_{50}) after exposure to cold for 2 hrs or more, or after exposure to heat for 0.5 hr. (From the data of numerous authors as cited in Larcher, 1973)

Plants	°C for cold injury in winter	°C for heat injury in summer
Tropics		
Ferns and herbaceous flowering plants	+5 to −2	45−48
Trees	+5 to −2	45−55
Subtropics		
Succulents		50−55
Sclerophyllous woody plants	−8 to −12	50−60
Temperate zone		
Evergreen woody plants of coastal regions with mild winters	−6 to −15	50−55
Dwarf shrubs of Atlantic heaths	ca. −20	45−50
Deciduous trees and shrubs with broad distribution	(−25 to −40)[a]	ca. 50
Herbs		
Sunny habitats	−10 to −20	48−52
Shady habitats		40−45
Water plants	ca. −10	38−42
Cold-winter areas		
Evergreen conifers	−40 or lower	44−50
Alpine dwarf shrubs	−20 to −70	48−54
Herbs of the high mountains and the arctic	(−30 to −196)[a]	44−54

[a] Leaf primordia in vegetative buds.

Constitutional Types with Respect to Cold

On the bases of the limits and the specific nature of cold resistance, one may distinguish three categories:

1. Chilling-Sensitive Plants. This group includes all plants seriously damaged even at temperatures above the freezing point: algae of warm oceans, some fungi and certain vascular plants of the tropical rain forests.

2. Freezing-Sensitive Plants. These can tolerate low temperatures, but they are damaged as soon as ice begins to form in the tissues. Freezing-sensitive plants are protected from injury only by mechanisms that delay freezing. In the cooler seasons the concentration of osmotically effective substances in the cell sap and in the proto-plasm is increased, which causes the freezing point of the tissue to be lowered to about $-12°$ C (in exceptional cases, to $-20°$ C). During the growing season all vascular plants are freezing-sensitive. Plants sensitive all year round include the benthic algae of cold oceans and some fresh-water algae, tropical and subtropical woody plants, and various species from warm-temperate regions (cf. also Table 37).

3. Frost-Tolerant Plants. In the cold season, frost-tolerant plants survive extracellu-lar freezing and the associated withdrawal of water from the cells. This category includes certain fresh-water algae and the algae of the tidal zone, but comprises mainly aerial algae, mosses of all climatic zones (even tropical), and the perennial land plants of regions with cold winters. Some algae, many lichens, and a number of woody plants can become hardened to extreme cold; they remain unharmed even after exposure to long periods of hard frost and can even be cooled to the tempera-ture of liquid nitrogen.

Constitutional Types with Respect to Heat

The effects of heat depend on the duration of exposure; that is, a slight excess of heat applied for a long time is as injurious as intensive heat applied for a short time. Therefore the heat-resistance data for plants must be standardized with respect to duration of exposure, and the generally agreed standard time is one-half hour. If the high temperature were instead maintained for an hour, the resistance limits would be about $1-2°$ C lower. The categories of heat resistance are as follows:

1. Heat-Sensitive Species. This group includes all species that are injured even at $30-40°$ C, or at the very most $45°$ C: eukaryotic algae and submersed vascular plants, lichens in a hydrated state (these, however, soon dry out in strong sunlight and then are completely heat-resistant), and most of the soft-leaved land plants. In addition, various bacteria pathogenic to plants, as well as viruses, are destroyed even at relatively low temperatures (for example, tomato wilt virus is killed at $40-45°$ C). All these species can colonize only habitats in which they are not exposed to over-heating—unless, of course, they are capable of keeping their own temperature down by means of transpiration (*Untertemperatur* species, in the terminology of O. L. Lange).

2. Relatively Heat-Tolerant Eukaryonts. The plants of sunny and dry habitats are, as a rule, capable of developing resistance to heat; they can survive heating for half an hour to $50-60°$ C. Just above $60°$ C there seems to be an absolute limit for survival of highly differentiated plant cells (i.e., those with nucleus and other orga-nelles).

3. Heat-Tolerant Prokaryonts. Some thermophilic prokaryonts can endure exceed-ingly high temperatures: bacteria, as much as $90°$ C, and blue-green algae as much as $75°$ C. These, like the heat-stable viruses, are equipped with especially resistant nucleic acids and proteins.

Temperature Resistance and Organ Function

Severe cold and extraordinary heat are events of limited duration, not uncommonly evaded by plants in that they discard sensitive structures or reduce their vegetative parts to subterranean organs designed to outlast such situations. When the unfavorable season has passed, they sprout anew. From an ecological point of view, one must know the resistance of the various plant organs to the temperatures they actually encounter, and take into account their functions.

Figs. 135 and 136 show the extent to which different organs, and even tissues in the same organ, vary in this respect. Particularly cold-sensitive organs are those for

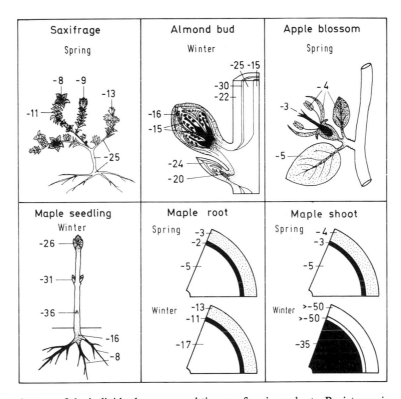

Fig. 135. Frost resistance of the individual organs and tissues of various plants. Resistance is given in °C. For sycamore-maple (*Acer pseudoplatanus*) seedlings the temperature given as the resistance limit is the lowest at which no injury is detectable (TL$_0$), but in all other examples it is the temperature for 50% damage (TL$_{50}$). The most sensitive tissue (or organ) is shown in black, and the most resistant, in white. After reproductive parts have been destroyed by freezing, the vegetative organs, as a rule, remain functional; and after extensive damage to the vegetative parts of a plant, particularly resistant tissues from which regeneration can occur usually still survive. If a plant is to maintain itself in habitats under year-round threat of frost (e.g., *Saxifraga oppositifolia* in the high mountains), it is advantageous for the regenerative parts to retain relatively high resistance even while the plant is actively developing. (From measurements by G. Rehner as cited by Pisek, 1958, and after Harrasser, 1969; Larcher, 1970; Kainmüller, 1974)

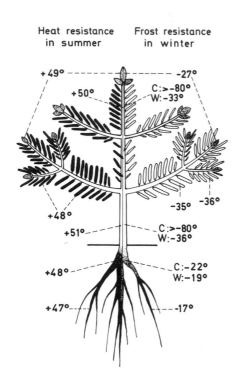

Heat resistance   Frost resistance
   in summer   in winter

+49° -------- ----- -27°
  +50°   C:>-80°
                     W:-33°

+48°
  +51°-------- C:>-80°
              W:-36°

+48°--------- C:-22°
              W:-19°

+47°---------- ---17°

             -35° -36°

Fig. 136. Heat resistance (in summer) and frost resistance (in winter) of various organs of a young fir tree. All temperatures given are those for 50% injury. The root is in each case the most sensitive organ and the shoot axis (*C* cambium, *W* wood) is the most resistant. The needles are less resistive to heat in summer than the buds, and in winter the buds are distinctly more sensitive to frost than are the needles. (After Bauer, 1970; Bendetta, 1972)

reproduction: the flower primordia in the winter buds and the ovaries within the flowers. In connection with questions of plant distribution in particular, the resistance of the floral primordia, the flowers, the seeds, and the unprotected new growth must be considered, for it is these most sensitive stages that limit fundamentally the maintenance and spread of a species (Thienemann's Rule). *Subterranean organs*, too, are relatively sensitive to heat and cold. In woody plants the resistance of the lignified parts of the root system—especially the root collar—is crucial in the survival of the whole plant; should these parts die off, the shoot must eventually perish. The *shoot* is the least sensitive to cold and heat. When hardened, its cambium has the greatest resistance of all tissues. An important consideration in estimating the consequences of temperature stress is the *bud resistance* as compared with the resistance of the leaves. Loss of leaves is not necessarily serious, as long as the buds remain healthy. Even if the buds are lost, however, some plants can produce new shoots from more resistant basal latent buds. Trees that have been forced to do this often become shrublike.

Differences in Resistance within Populations

In a given population, plants of different *ages* are not uniformly resistant, nor are they uniformly endangered. This situation is illustrated in Fig. 137, as exemplified by a stand of Mediterranean evergreen oaks. Lowering of the soil surface temperature to only −4° C can destroy all the potential seedlings from that year; if the temperature falls to between −8° and −10° C in several successive winters, a natural regen-

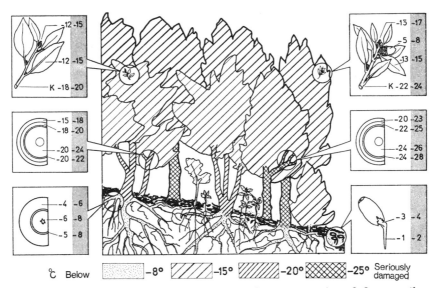

°C Below ░░░ -8° ▨▨ -15° ▧▧ -20° ▩▩ -25° Seriously damaged

Fig. 137. Zonation of frost-resistance levels in winter in a community of *Quercus ilex*. Temperatures indicated in the unshaded part of the marginal pictures represent the extremes below which first frost damage is to be expected; the data in the shaded parts refer to 50% damage. "K" in the upper pictures denotes the resistance of the cambium. Young plants, from one to three years old, are killed by temperatures between −10° and −15° C. (From Larcher and Mair, 1969)

eration of the community becomes impossible, even though the mature shrubs and trees suffer not the slightest injury at these temperatures. Only temperatures as low as −20° to −25° C are catastrophic for such a stand of oaks, and then only if the cold spells are long enough to freeze the thicker trunks. It is advantageous to the stand that the outer parts of the crowns, the parts most exposed to frost (radiation frost), develop a greater resistance to cold than the young stages growing up in the shade of the stand. The various components of a forest are adapted to the degree of cold normally encountered in their local subhabitat. This is also true of plants in the shrub and herbaceous strata; their shoots and perennating buds are necessarily more resistant to cold, the further they project above the soil and the layer of litter (Table 40).

Not only the different age groups, but even the different *individuals* of a population vary with respect to the nature and degree of thermal resistance. This diversity can be the basis of the evolution of especially resistant ecotypes and varieties, which survive even though the climate eventually becomes so unfavorable that the continued existence of the species in that area is threatened.

Prospects of Survival and the Limits for Existence

The probability that a plant species will survive is a function of its ability to come unscathed through extreme weather conditions and to maintain itself in endangered habitats. During a heat spell, the plants must cope not only with high temperature

217

Table 40. Frost resistance of leaves, buds and subterranean organs, in various life forms. (From Till, 1956)

Plant part	Frost resistance in winter (°C)
A. Subterranean organs	
Roots, rhizomes, perennating buds	− 6 to −13.5
B. Organs above ground	
1. Herbs (Hemicryptophytes)	
a) Winter leaves	
3—5 cm above the litter layer	−11.5 to −14.5
5—10 cm above the litter	−11.5 to −18.0
10—20 cm above the litter	−13.0 to −20.0
b) Buds	
under leaf litter	− 7.0 to −11.5
just above the litter	−12.5 to −18.0
3—20 cm above the litter	−15.5 to −19.5
2. Shrubs and dwarf shrubs (Chamaephytes)	
Sarothamnus scoparius, assimilating axes	−18.5
Erica tetralix, leaves	−20.0
Erica tetralix, buds	−19.5
3. Trees (Phanerophytes)	
Fagus sylvatica, leaf buds	−29.0
Betula pendula, flower buds	−38.0
Betula pendula, leaf buds	−40.0

but also with the threat of desiccation; winter brings not only the direct effects of cold but also such dangers as frost drought, the pressure of snow, and avalanches. The *probability of survival* of a species in a given habitat is higher, the greater the resistance of the most vulnerable vital part of the plant to all these factors, the sooner injuries are healed (the recovery capacity), and the less frequently the extreme conditions recur.

The *recovery capacity* of a plant is not easy to ascertain. The likelihood that damage can be repaired is more readily estimated if the resistance of the perennating buds and associated structures is known, and if the occurrence of unusually long and severe winters, late or early frosts, and abnormally hot spells is studied. In particular, it is advantageous if these climatic excursions can be employed as "field experiments", to test existing concepts in the natural habitat.

In order to judge the degree to which a plant is *endangered by stress*, one must have access to data about distribution, frequency, and probable time of occurrence of extreme temperatures. When suitable measurements of plant temperature are not available, temperature data from standard weather stations can provide rough estimates for large-scale comparisons involving plants of adequate size. Such inferences may be useful, but one must keep in mind that they are no substitute for a full, quantitative ecophysiological analysis.

Periodicity in Climate and Vegetation

The rate of absorption of radiant energy at various places on the earth's surface depends upon their orientation with respect to the sun. This direction-dependence, because of the rotation and revolution of the earth, makes energy input a periodically varying environmental factor. It imposes periodicity upon all terrestrial phenomena, including the lives of organisms.

Climatic Rhythms

Diurnal Variation

The earth's rotation causes an alternation between day and night, between hours of light and of darkness; responses of plants to this alternation are termed *photoperiodism*. Responses to the corresponding temperature alternation reflect *thermoperiodism*. The changes in temperature, as a rule, lag somewhat behind those of light; the air temperature reaches a maximum not at the time the sun is highest but a little later, and the daily minimum temperature is reached near the end of the night.

At latitudes near the equator the photoperiod changes little in the course of the year. At the Tropics of Cancer and Capricorn the difference between the longest and shortest days is only 2 hrs. Thus there are no well-defined thermic seasons in the tropics, and the daily cycle is the important climatic rhythm (Fig. 138: Quito).

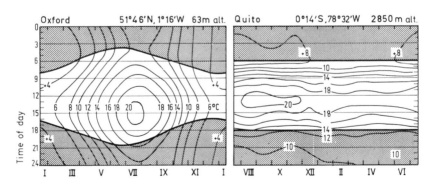

Fig. 138. Average air temperature, as a function of both time of day (ordinate) and time of the year (abscissa, Roman numerals for the months). Isotherms connect loci of constant temperature, and one can read the daily temperature variations by scanning vertically at any time of the year. Duration of daylight and nighttime (shaded areas) hours are also shown. Left: *Oxford*, England, is at an intermediate latitude and has a maritime climate (slight daily temperature fluctuations). Right: *Quito*, Ecuador, has an equatorial highland climate (marked daily fluctuation of temperature with almost no annual fluctuation). (After Troll, 1955)

Where there is strong insolation, on high plateaus and in the mountains, the daily cycle can be marked by very pronounced temperature oscillations. In tropical mountains the plants at high altitudes are exposed to night frosts (down to $-5°$ C) all year round, and during the day the temperature near the ground rises to $30°$ C or more.

Seasonal Variation

At higher latitudes the lengths of day and night change increasingly in the course of a year, and beyond the polar circles alternation between day and night ceases altogether in periods centered about the solstices. At an intermediate latitude such as that at Oxford (Fig. 138), the day lasts about 16 hrs in summer and only half as long in winter. This seasonal variation in photoperiod is accompanied by seasonal changes in average temperature. Toward winter the radiation balance becomes negative, primarily because of the steadily increasing duration of the nighttime phase in which reradiation predominates, but also because of the lowered intensity of insolation at the low sun elevations. The relatively dark season thus becomes a cold season.

Vegetative processes may be suppressed not only by a season in which light is limited and temperature low, but also by periodic seasons of dryness. These are brought about by seasonally determined shifts in large-scale atmospheric circulation, and these in turn are caused by the differences between the energy balance in the strongly irradiated lower latitudes and that in less strongly irradiated northerly and southerly regions.

Activity Rhythms

Growing Season and Climatic Rhythmicity

As a result of the periodicity of insolation, day length, temperature and precipitation, there is a regular alternation between times when conditions are favorable to plant growth and those when they are not. Plants adjust to this climatic rhythmicity through rhythmic changes in the state (permeability, viscosity) of their protoplasm, in their metabolic activity, in their developmental processes, and in their resistance to environmental stress. Their life cycles are adapted primarily to the duration of the growing season and to the conditions prevailing during that season. North of $40°$ latitude the days are longer than the nights during the entire growing season, and above $50°$ latitude the difference is quite pronounced (Fig. 139). If a species is to thrive and extend its range in a certain region, it must be able to coordinate its life cycle with the progression of the periodic phenomena in its environment. A plant species, a variety, or an ecotype is well acclimatized if the growing season is utilized to the full without risk of injury as the unfavorable season approaches. In general, this is ensured by the fact that the acquisition of resistance is coupled to developmental processes (cf. Fig. 133). Poorly adapted plants might sprout too late, continue to develop too slowly, and be damaged by the first winter frosts; conversely, the situation would be equally disadvantageous if they began to grow too

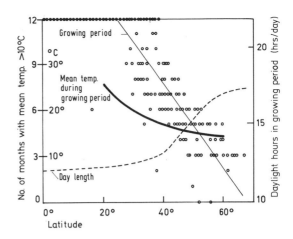

Fig. 139. Duration of the growing season (number of months with mean air temperature above 10° C) in locations at different latitudes (circles), and the dependence of day length and the mean temperature during the growing period upon latitude. At low latitudes the temperatures are favorable for growth and the days are short all or most of the year. At high latitudes the growing season is restricted to a short time in the summer, but then the days are long and the air temperature is on the average not much lower than during the growing season at intermediate latitudes. (After Totsuka, 1963)

early in the year (risking injury by late frosts) and stopped development too soon to make full use of the favorable season (Fig. 140). A lack of synchronization between periodic plant activity and the rhythmicity of the climate thus restricts the spread of a species; such maladaptations can be overcome, of course, by the evolution of better adapted races.

The Seasonal Course of Growth and Development

Growth and development proceed according to a genetically set norm, coordinated by hormones and modified by environmental influences. In considering the interaction of these internal and external factors, it is helpful to subdivide the development of a plant into certain phases; these may merge gradually or may take the form of distinct successive stages.

The Succession of Phases in a Life Cycle

The life of any organism begins with a reproductive process; this is followed by vegetative developmental processes such as growth and the formation of organs, which in turn are followed by the reproductive processes leading to the next generation, which complete the cycle.

Life Cycle of Annual Plants. In annual plants the foregoing phases of the life cycle follow one another in uninterrupted sequence. Immediately after germination, the vegetative organs acquire their definitive form and multiply in number. When the intensive growth of the shoot subsides, the plants flower. Even while the fruits are ripening, signs of aging appear in the vegetative parts of the plant—the breakdown of proteins and the yellowing of the leaves. Finally the whole plant dies, leaving only the seeds; these remain in a dormant state until they are aroused by conditions favorable for germination.

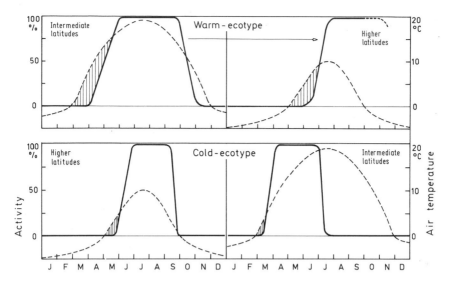

Fig. 140. Diagram illustrating the temporal coordination of the climatic rhythm and the rhythm of growth in trees. The *left half* of the figure shows a general synchronization between progression of temperature through the year (dashed curve) and physiological activity (solid curve); in the *right half* the consequences of lack of adaptation are indicated. Ecotypes adapted to warm conditions do not send out shoots until the temperature has risen considerably (cf. the areas shaded with widely spaced lines), whereas ecotypes from higher latitudes, adapted to cold, end their winter dormancy as soon as there is a slight rise in temperature (area with closely spaced lines). If warm-climate-ecotypes are introduced to high latitudes (upper right), they retain their tendency to delay the onset of development; they sprout too late and the new growth has no time to mature. Thus they can be caught by frost before they have entered winter dormancy and may be damaged. Cold-climate-ecotypes transplanted to a warmer region (lower right) sprout prematurely and therefore are endangered by spring frosts; moreover, they conclude their growth—which is adjusted to a short growing season—much too soon and fail to make use of a large part of the favorable season. (From transplanting experiments by Langlet as cited by Bünning, 1953)

Perennial plants. In perennials, especially woody plants, development proceeds more slowly and in repeated steps. Each step of vegetative development begins with the sprouting of buds and the expansion of leaves, and ends with the maturation of the new growth, the forming of new buds, and the abscission of leaves (in evergreens, the leaves discarded are from previous years). When the plant has reached the age of flowering, the events in each period of growth are supplemented by those of the reproductive cycle (from formation of floral primordia to the falling of the fruits). Depending on the species, vegetative growth and reproductive development may occur either simultaneously or in alternation. Simultaneity occurs especially in the tropics; typical examples are the coconut palm, banana and mango tree. Examples of alternation between vegetative and reproductive phases are the biennial shrubs and some geophytes (for example, meadow saffron). In many species the formation of the flowers (though the primordia are laid down in the growing season) is such a long-drawn-out process that flowering occurs only after the leaves have fallen. In tropical and subtropical dry regions, especially, there are woody plants that bear their flowers on bare twigs at the end of the growing season. In some cases the fully differentiated floral primordia are inhibited from further development even during the subsequent dormant season; such plants are of the early flowering type and include many fruit trees and those arctic and high-mountain plants that open the flowers formed in the previous year soon after the spring thaw, even before the leaves have appeared.

Phenology

In regions with alternating seasons, the life cycle of the plants is synchronized to the long-term changes in weather. The onset and duration of different developmental phases are thus different from year to year. Phenology is the study of the cycle of

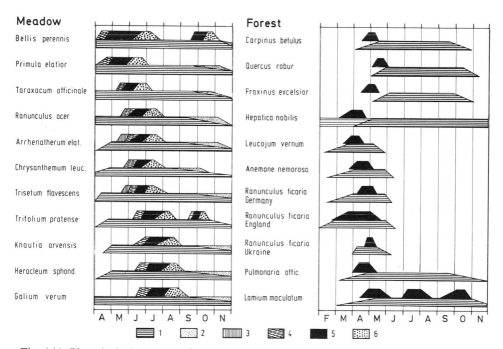

Fig. 141. Phenological spectrum for an oat-grass association (Arrhenatheretum elatioris) in Poland (left) and for species of the tree and herb strata of a mixed forest dominated by oak in northwestern Germany (right). In the latter association, graphs for *Ranunculus ficaria* from England and the Ukraine are also given for comparison. The representation of phenological development by two horizontal bands is taken from Schennikow; in the lower band vegetative development is indicated and in the upper, reproductive development. *1* leaf-bearing period; *2* yellowing of leaves; *3* production of grass culms; *4* formation of flower buds; *5* blossoming period; *6* ripening of fruit and dispersal of seeds. For *Hepatica* the transition from foliage of the previous year to newly formed foliage is indicated by the different spacing of the lines. (After Salisbury, 1916; Ellenberg, 1939; Jankowska as cited by Lieth, 1970; Goryshina, 1972)

(A) Northern prealpine region

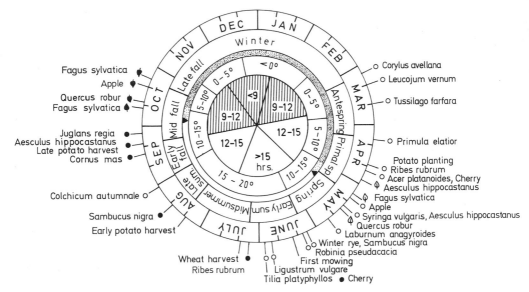

Fig. 142 A and B. Phenological calendar for the northern and southern prealpine lowlands. *Circles*: onset of flowering; *light leaf symbol*: expansion of leaves; *dots*: onset of fruit-ripening; *dark upright leaf symbol*: discoloration of leaves; *dark inverted leaf symbol*: leaf fall. *Rings of the circular calendar, from the outside in*: months, mean duration of the phenological seasons, period of subfreezing temperatures (dotted), average period with mean temperature in the indicated range; *central circle*: duration of daylight. A close comparison of figure A with figure B reveals characteristic differences in the times of onset and rearrangements in the sequence of certain phenological dates. For example, the Norway maple flowers before the apple trees north of the Alps, whereas south of the Alps they flower together. Such shifts in general are termed "phenological interception". One can also learn from the diagrams whether the flowers and the young shoots of the different species of plants (e.g., apple) in a given area are endangered by late frosts or not. (After Larcher, 1964, with reference to phenological data in Schnelle, 1955; Marcello, 1959)

sprouting, blooming, bearing of fruit and senescence (as well as other biological phenomena with annual periodicity), as they are affected by climatic changes. The science of phenology was founded by C. Linnaeus as early as the mid-18th century, but phenological observations had of course been made, and given practical application, much earlier. For centuries in Japan, the time of the first flowering of the cherry trees has been considered significant, and much of the traditional wisdom of farmers demonstrates a capacity for sharp observation and a deep insight into the relationship between the progression of meteorological phenomena and the development of the vegetation.

Phenological Dates. Phenology is based, even today, on the observation of externally visible changes (*phenophases*) in the course of a plant's life cycle. Phenological descriptions provide ecologically valuable information about the average duration of

(B) Southern prealpine region

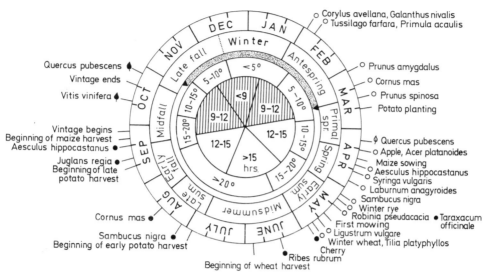

Fig. 142 B

the growing and foliated periods of the plant species in an area (Fig. 141), and about local and weather-determined differences in the dates of onset of such phenomena (Figs. 142 and 143). As a science of the interrelations between biology and climatology, phenology is not a simple task, since the environmental factor eliciting an event (for example, the passing of a critical temperature threshold) is long past by the time the associated phenophase becomes apparent. Nevertheless, even simple observations suffice for the recognition of certain relationships:

1. The time of onset of *phenophases of the first half-year* depends primarily on the passing of certain *temperature thresholds*. This can be shown by comparing the temperature distribution over a certain terrain with the phenological dates (Figs. 143 and 144). The opening of the buds, sprouting, the onset of flowering in trees and shrubs and the germination of seeds are possible only after the temperatures of both air and soil exceed regularly a critical point characteristic of each stage. In general, the temperature threshold for the opening of the buds and flowering is 6–10° C, though in early-flowering and mountain plants it is lower (in some cases around 0° C) and in late-flowering plants, higher (in many ring-porous trees between 10° and 15° C, and about 15° C in grain). Poplars, birches and some species of conifers sprout at just above 0° C. Sprouting and flowering, however, can be elicited by warmth only if the plants are already in a state of readiness to develop—i.e., if they hav emergD from their winter dormant period (cf. p. 232).

2. Phenological dates falling in the *second half-year*, such as those for ripening of fruit, discoloration of leaves, leaf fall, and the times for harvest of crops, can be affected by all those environmental conditions that delay or accelerate the processes of maturation and aging. Once again, temperature is of greatest significance, but in

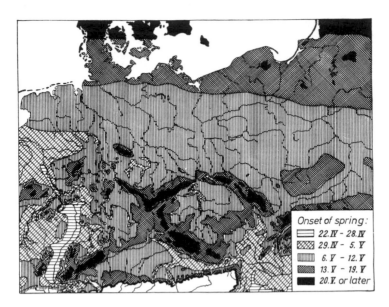

Fig. 143. Map of the flowering period of the apple tree. Flowering of the apple is characteristic of the onset of spring. In the northern part of this region the onset of spring progresses by zones, whereas in the mountainous south the progression is in term of altitude, from the valleys up the slopes. (After Ihne as cited by Walter, 1960)

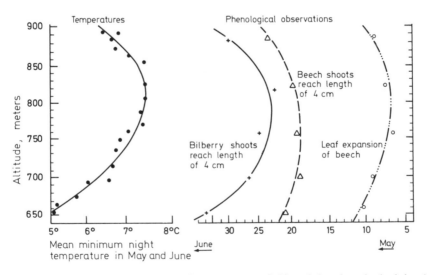

Fig. 144. Relationship between night temperature (left) and the phenological development of forest plants (right) at different altitudes on a mountain slope in southern Germany. (After Frenzel and Fischer as cited by Geiger, 1961)

226

this case its role is evident with respect to the enhancement of synthetic activity. Thus threshold temperatures are of less concern than the *total heat*—i.e., the number of hours in which the temperature has exceeded given levels. Other decisive factors are the supplies of nutrients and water and, above all, the influence of the diurnal photoperiod upon the times of onset of flowering, leaf fall, and winter dormancy. In various species the plant is prepared for discoloration and leaf abscission by the shortening days. Then, as soon as the temperatures fall below threshold values between 5° and 10° C, these closing phases of the phenological calendar appear. Regional variations of these dates are also of interest; for example, in mountains the falling of leaves begins at high altitudes and progresses rapidly toward the valleys.

Phenometry

For a detailed analysis of the influence of external factors upon the course of development, simple phenological observations of dates of onset of the various phases are not enough. It is necessary to have phenometric records of growth, particularly with respect to gradual processes like the expansion of leaves, elongation of the shoots and roots (Fig. 148), and increase in thickness of the cambium. Phenometry also includes studies of the increase in mass and leaf area of individuals and stands of plants (Fig. 145). A good example of a problem in this field is that of cambial growth.

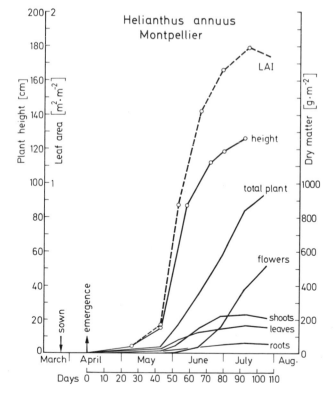

Fig. 145. Phenometric analysis of the growth of sunflowers in southern France. The expansion and die-off of the leaves is expressed quantitatively by the leaf area index (*LAI*), and the growth of shoot and roots, by the dry matter production. Note that the different organs do not all develop in parallel. (After Eckardt *et al.*, 1971)

227

Phenometry of Annual Growth Rings. The level of mitotic activity in the cambium is closely related to the increase in length of the shoot. In annual plants cambial activity begins when elongation is to a great extent completed; in woody plants it is initiated each year in connection with the sprouting of new shoots and the unfolding of the leaves. The first sign of beginning activity is a swelling of the cambium initials, in which the radial walls are stretched out and can be easily torn. Correspondingly, the phenological signal of the beginning of cambial activity is that the twigs can be peeled (the bark is easily detachable). Soon thereafter the first divisions occur. In many trees of the tropics and subtropics, as well as in the fig, olive and the seedlings of some woody plants, growth in diameter is initiated, and stops, quite suddenly whenever external conditions happen to be appropriate (Fig. 146). In the woody plants of temperate latitudes just one growth ring is normally formed per year; hence the term *annual ring*. The ring is composed of histologically distinguishable zones in the wood, corresponding to the spring and the late summer growth.

The duration of cambial activity and the type of wood formed—the differentiation into early or late wood—are affected by environmental factors. Precise measurements of the increase in thickness of the stems during the year, using "dendrometers", and histological analyses of the new wood thus provide informative phenometric data. In general, the formation of spring wood is promoted by all those factors that favor the sprouting of buds and growth in length of the new shoots; all factors tending to slow shoot growth and accelerate the aging of the foliage lead to differentiation of late wood. The thickness of the cell walls in the new wood depends upon

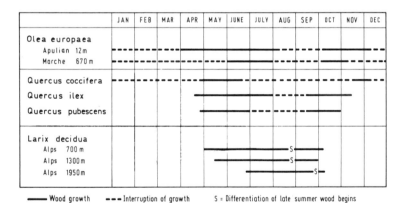

—— Wood growth   - - - Interruption of growth   S = Differentiation of late summer wood begins

Fig. 146. Seasonal periodicity of cambial activity and growth of new wood in different species of woody plants and in trees of the same species in climatically different habitats. In the olive tree and in the evergreen oak *Quercus coccifera*, increase in diameter and the differentiation of xylem cease as a *direct* effect of both dryness in summer and falling temperatures in winter; the interrupted growth is resumed immediately when conditions again become favorable. The two other oak species (*Q. ilex* evergreen, *Q. pubescens* deciduous) interrupt cambial activity in summer and conclude the annual growing season on the basis of an *endogenous rhythm*. The example of the larches—from sites in the valley, on the slope and at the timberline—shows that as altitude increases cambial activity begins later and the differentiation of late wood is stopped prematurely. (After Messeri, 1951; Tranquillini and Unterholzner, 1968)

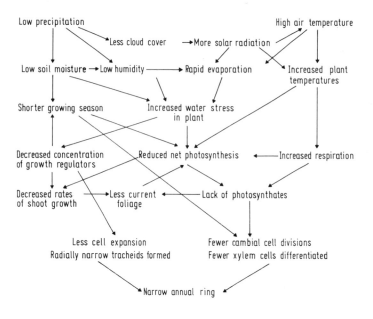

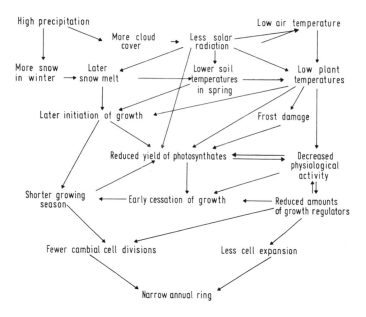

Fig. 147. Model of the causal relationships between different environmental conditions (precipitation and temperature) and the size of the annual rings of a conifer. An abnormally narrow ring can be the result either of inadequate precipitation and high air temperature (above) or of heavy precipitation and low air temperature (below). The model indicates the diversity of interrelationships, and can usefully be studied in detail both from top to bottom and in the reverse direction (against the arrows). (After Fritts *et al.*, 1971)

the supply of carbohydrates, and thus is an indication of the yield of synthetic metabolism in that year. Other factors having direct or indirect influences on the thickness and appearance of the annual rings include radiation, temperature, availability of nutrients, water supply and duration of the photoperiod, as well as all kinds of harmful environmental influences; examples of the latter are attack by parasites, consumption by animals, excessive heat and frost, and absorption of pollutants. In cases where the particular influences to which the cambium is most subject are known in detail for a given species, the structure of the annual rings provides an important historical document of the growth-determining events of past years (cf. Fig. 147).

The Alternation of Periods of Vegetative Activity and Dormancy

The regular alternation between developmental activity and periods in which growth is temporarily slowed or stopped can be imposed on the plants by the recurrence of unfavorable environmental conditions. But the process can also, to a considerable degree, be preprogrammed and occur spontaneously. According to W. Pfeffer, it is useful to distinguish between *aitionomic* (imposed by the environment) interruptions in growth and *autonomic* (innate) rhythmicity having its origin in the genotype.

Aitionomic Alternation between Activity and Suspended Growth

Some plants have no provision for rest periods in their schedule of development. Their life cycles proceed without interruption unless they are forced to pause by unfavorable external factors. Many winter annuals (e.g., *Senecio, Cerastium, Capsella*) cease development only when the temperature falls below the freezing point, and resume growth when there is a thaw. The cryptogams, too, to the level of the ferns, suspend their growth only temporarily when forced to do so by external conditions and resume growth when the situation improves. In perennial flowering plants continuous *shoot* growth is rare. The growth of the *roots*, on the other hand, which in many respects are the most primitive plant organ, is evidently only aitionomically regulated. The most decisive factor is soil temperature; lack of water affects growth only when the deficiency is very serious. The temperature range over which elongation of the roots can occur is usually very wide, and the minimal limiting temperature for root growth in woody plants of the temperate zone is rather low, between 2° and 5° C. It is thus not surprising that roots begin to grow before the buds sprout and that they continue growing until late in the fall (Fig. 148). Plants of warmer regions require higher temperatures. *Citrus* roots grow only above 10° C; within the natural geographic range of this genus, the soil temperature never falls below this point, even during the coldest season. But for orchards, this temperature limit can be critical, since it is in winter that the fruits of most *Citrus* species ripen—a process requiring a good supply of water and nutrients from the roots.

Autonomic Rhythmicity

Most flowering plants have an inherent, genetically determined tendency (hence one also speaks of "*endogenous*" or innate rhythmicity) to alternate between activity and

230

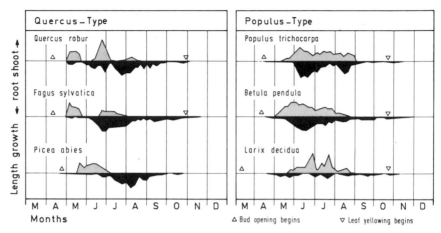

Fig. 148. Growth in length of shoot (increasing upward and stippled) and of roots (increasing downward and black) in several species of tree during the course of the year. In the *oak type* the growth of shoots is concluded early in the year; it proceeds in successive spurts separated by endogenously fixed growth pauses. Usually there are two such spurts per year, but the second is sometimes omitted; often there is an additional growth of shoots in late summer. This type includes pine and fir as well as the species shown in the figure. In the *poplar type*, shoot growth is synchronized by day length and the climatic temperature changes. This type also includes linden and black locust. The *growth of roots* in both types is regulated primarily by external factors; frequently it begins before shoot growth and continues until late in the autumn. (After Hoffmann, 1972)

rest. This finds expression in the fact that the growth in length and thickness of the shoot occurs in spurts rather than continuously, as does the changing of the leaves, and the rhythmical filling and emptying of storage organs. The defining property of autonomic rhythmicity is its existence and continuation without external synchronizing factors. Even under the conditions prevailing in permanently humid tropical zones, which are favorable for plant growth all the year round, only 20% of all evergreen trees grow continually like annual herbs; the remainder pause for certain periods during their development. Grasses and geophytes retain their natural rhythm, evolved as an adaptation to the conditions of temperature and precipitation in the steppes and savannas, even when they have spread into regions where such behavior would not be necessary.

In regions with climatic rhythmicity, the genetically based "physiological clock" is adapted to the regular fluctuations in meteorological conditions (see, for example, the poplar type in Fig. 148). But even in cases of apparent synchronization, a close analysis of shoot growth discloses a pronounced autonomy of *some* species. A prime example of such endogenous periodicity is given by beeches and oaks, the shoots of which cease to elongate early in the summer—at a time when the days have not yet begun to shorten and other external factors also present no hindrance to growth. Some weeks later the plants become ready to resume development, and there may be a renewed shooting ("lammas" shoots). Not until autumn weather conditions prevail is growth brought to a final halt (Fig. 148, oak type).

The Winter Rest Period of Woody Plants

Toward the end of the summer, lateral buds are formed in the axils of the leaves, and the tips of the shoots become transformed into winter buds, or wither and die off. When the foliage begins to yellow, these buds are already dormant. Other parts of the plant also enter dormancy for the winter; the cambium, for example, and the other tissues of the shoot, can thus become hardened to frost and dehydration. Reserve substances (starches and fats) are deposited, various metabolites and minerals are shifted, and the complement of enzymes is altered: hydrolases and catalases increase in the fall, polyphenoloxidases decrease, and related „isoenzymes" appear in the place of a number of enzymes; the actions of the isoenzymes differ from those of their counterparts only with respect to temperature dependence. This readjustment to the winter state is not an abrupt event, but occurs gradually, some alterations appearing earlier and others later; the transformation is not synchronized throughout the plant.

Distinct phases are observable in cases of endogenous winter dormancy. In fact, as W. Johannsen has noted, there are in general three successive stages of dormancy: predormancy, true dormancy and postdormancy (Fig. 149).

1. Predormancy. This stage begins in the buds even before leaf abscission. In some woody plants, reduced day length induces the conclusion of the growth period and the transition to the dormant state; this is the case, for example, in poplar and willow, birch, hazel, beech, oak, maple, spruce and larch, most of them species with distribution extending far to the north. The *critical day length* is around 12—15 hrs; in subarctic ecotypes of spruce it can be as great as 20 hrs. Therefore, in connection

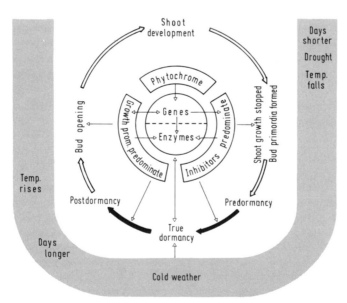

Fig. 149. Simplified diagram of environmental influences (shaded U), and hormonal interactions with cell activity and the developmental rhythm, in woody plants

with induction of dormancy, one should speak not of "short" days but rather of the days growing shorter. Very long days prevent the onset of winter dormancy in all these species, as well as in some species of juniper and many Atlantic plants. Long days can, moreover also initiate the transition to bud dormancy—not in woody plants, but in geophytes such as lily-of-the-valley and various bulb plants. Species responding in this way evidently have *two* built-in rest periods, one at the time of greatest winter cold and the other during the summer drought.

The second important regulating factor is *low night temperature*; critical levels are usually a little under 10° C, depending on the species. In all species not made dormant endogenously or by decreasing day length, falling temperature is the decisive signal. These are primarily genera native to the southern temperate zone, such as ash, horse chestnut, lilac, cherry and many others. Falling temperatures are apparently more effective in general than short days, for they can replace the latter in experiments upon many photoperiodically controlled plants. In nature, photoperiod and thermoperiod are necessarily coupled, so it is not surprising that growth rhythms and many other vital processes are controlled by the joint action of these two variables. It is not uncommon to find that change in day length acts as the pacemaker inducing the transition in the plant, the process then being completed under the influence of changes in temperature.

2. True Dormancy. During predormancy, inhibition of bud activity increases steadily until finally, in November and December in the temperate latitudes, complete dormancy occurs. At this point the plants can no longer be induced to sprout by warming or by lengthening of the photoperiod. The inability of plants to emerge prematurely from the resting state once they have entered true dormancy is an important factor in their resistance, in view of the unpredictability of the weather. The plants of cold-winter regions must not respond to mild days in midwinter, or they would unavoidably suffer injury in the next cold wave. In species lacking a deep winter dormancy (*Citrus* species), this in fact happens time and again. Still lower temperatures accelerate the transition to postdormancy. Many woody plants, such as poplar, maple, linden, pine, some fruit trees and the grapevine have a distinct requirement for cold; they can sprout normally only after a certain exposure to low temperatures—usually to *ca.* 0° C for 3—4 weeks.

3. Postdormancy. Once true dormancy has been broken, treatment with heat (warm baths) and additional light cause the buds to develop. In intermediate latitudes, postdormancy usually ends in February. From that time on, environmental factors alone determine the time of sprouting.

Plant Rhythmicity and Phytohormones

Seasonal changes in activity, and the molecular mechanisms underlying them, are the subject of developmental physiology; findings in this field are of the greatest significance to ecologists. In particular, external factors cause an important shift in the relationship between the phytohormones that promote growth and those that inhibit it (Fig. 150). Chief among these changes is an apparent disappearance of gibberellic acid at the transition to dormancy, so that inhibitory substances such as

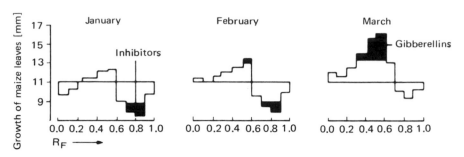

Fig. 150. Effect of chilling on the content of gibberellins and inhibitors in buds of the European black currant (*Ribes nigrum*). The growth regulators were separated by paper chromatography, and assayed by testing their effect on the growth of leaves. (After El-Antably and Wareing as cited by Wareing and Phillips, 1970)

abscisic acid dominate. Conversely, the concentration of the growth-promoting regulators is increased to some extent during predormancy and more prominently during postdormancy; metabolism is accelerated in preparation for development. The phytohormones serve to mediate between the influences of the environment and the genetically fixed mechanisms of synchronization. They act not only to coordinate growth and development, but also can modulate CO_2 fixation and the forms in which it is assimilated, water balance, and resistance to climatic extremes—thus adapting the intricate processes of plant life to the complex demands of the environment.

Synopsis

The goal of ecology is a comprehensive view of the diverse and numerous interactions of organisms and their environments, in their full complexity and ranges of variability. In interpreting ecological systems, the ecologist must try to understand the dynamic interplay of environmental factors, the reactions and adaptations of the organisms, and the regulatory mechanisms underlying them. All environmental influences, of course, act continuously and jointly upon the plant, which must adjust to them if it is to maintain itself in the face of competition. The study of these is the task of ecophysiology. An illustrative example of the complex interrelationships of environmental influences, metabolic balance and the processes of development is given in Fig. 147. Here one can see how a single measurable response of a plant (in this example, a narrow annual growth ring in the wood) can be brought about by a great variety of circumstances, and also how a given external factor can have extremely varied effects.

Special Characteristics of Ecological and Ecophysiological Methodology

Because of the complexity of environmental influences, and of the plant's reaction to them, research projects must involve cautious, long-term planning and the most comprehensive possible recording and evaluation of data. The model of the progress of an ecological research program given in Fig. 151 is designed to make clear the scope of the subject and the variety of methods employed in such a study.

As a first step, one must select the methods of measurement most suited to the object of the investigation and to the peculiarities of the site of the study. Not uncommonly, it turns out that some measurement devices must be newly developed for the purpose. The measurements themselves should include the spatial distribution and the temporal course of habitat factors and the physiological behavior of the plants; these should be continued for a long period, at least for a whole growing season. Automatic data-recording equipment should thus be used wherever possible (Fig. 152). Ecophysiological measurements in the field can be supplemented by related laboratory experiments. The latter yield reproducible data concerning the physiological behavior of the plants under a variety of environmental conditions, with the added advantage that the conditions of interest can be varied at will, singly or in groups. Laboratory measurements facilitate the understanding of field studies, and are particularly suitable for determining limiting factors.

Once an ecosystem has been analyzed in terms of the factors of interest, the next step is a "resynthesis" of the data so that the effects and interactions of the variables can be understood. With the large quantities of data ideally obtained, such evaluative procedures are best carried out by computer and their results presented in the form of diagrams (e.g., the summary of complex interactions in Fig. 147). The resulting models of subunits of the systems under study—individual plants or selec-

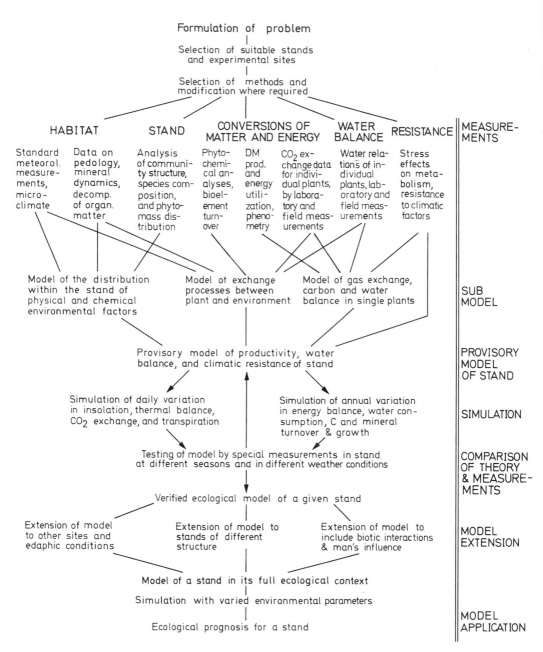

Fig. 151. Flow diagram of a program of research in experimental ecology, from the formulation of the problem to the ecological prognosis. (Based in part on Larcher *et al.*, 1973; Cernusca, 1973)

Fig. 152. Daily course of CO_2 exchange, transpiration and diffusion resistance in leaves of the grapevine as compared with the environmental conditions in an extremely dry habitat—an example of an ecological experiment including simultaneous measurement of several critical variables. I illuminance; F_n net photosynthesis; R respiration; E transpiration; Δ_e difference in water-vapor concentration between leaf and surrounding air (mg $H_2O \cdot l^{-1}$); r diffusion resistance for water vapor (sec $\cdot$ cm^{-1}); T_L leaf temperature; T_A air temperature; RH relative humidity in %. (From Schulze et al., 1972)

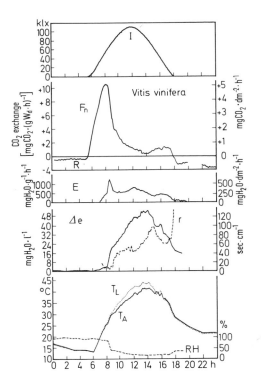

ted aspects of the environment such as radiation balance or gas exchange—are eventually combined to form a model of an entire stand of plants. In testing such models, computer simulations may be of help in suggesting new experiments or field observations.

When a model has proved to be a useful description of a restricted ecosystem, it can facilitate further study of the system. For example, the spatial range over which the model applies can be tested in a time- and labor-saving way by carefully planned short-term measurements under critical weather conditions. The ultimate goal, of course, is an *overall ecological model*, permitting mathematical analysis—for example, in terms of systems theory. The ecological predictions of such a comprehensive formalism will offer an objective basis for the critical decisions involved in planning for the future.

References

The publications listed here are those which have provided data and other original material for the figures and tables in this book. With few exceptions, the figures and tables have been modified.

ABEL, W. O.: Österr. Akad. Wiss., Math.-naturw. Kl., Sitzber. Abt. I, **165,** 619—707 (1956).

ALEXANDROV, V. Y.: Quart. Rev. Biol. **39,** 35—77 (1964).

ALTMAN, L., DITTMER, S.: Federation Am. Soc. Exp. Biol. Bethesda, Maryland **1966,** 60—121.

ARVIDSSON, I.: Austrocknungs- und Dürreresistenzverhältnisse einiger Repräsentanten öländischer Pflanzenvereine nebst Bemerkungen über Wasserabsorption durch oberirdische Organe, Suppl. 1, pp. 5—181. Kopenhagen: Oikos 1951.

AUBERT, B.: Oecol. Plant. **6,** 25—34 (1971).

AULITZKY, H.: Arch. Meteorol. Geophys. u. Bioklimatol., Ser. B **10,** 445—532 (1961).

AULITZKY, H.: Arch. Meteorol. Geophys. u. Bioklimatol., Ser. B **11,** 301—362 (1962).

AULITZKY, H.: Centr. ges. Forstw. **85,** 2—32 (1968).

BAEUMER, K.: Allgemeiner Pflanzenbau. Stuttgart: E. Ulmer 1971.

BARRS, H. D.: C.S.I.R.O., Div. Irrig. Res. Ann. Rep. **7,** 1968—1969 (1969).

BAUER, H.: Diss. Innsbruck 1970.

BAUER, H., HUTER, M., LARCHER, W.: Ber. deut. botan. Ges. **82,** 65—70 (1969).

BAZILEVICH, N. I., RODIN, L. Y.: Geographical regularities in productivity and the circulation of chemical elements in the earth's main vegetation types. In: Soviet geography (Rev. & Translation). New York: American Geogr. Soc. 1971.

BAZILEVICH, N. I., RODIN. L. Y., ROZOV, N. N.: Geographical aspects of biological productivity. In: Soviet geography (Rev. & Translation). New York: American Geogr. Soc. 1971.

BENDETTA, G.: Diss Innsbruck 1972.

BERGER-LANDEFELDT, U.: Ber. deut. botan. Ges. **71,** 22—33 (1958).

BERGER-LANDEFELDT, U.: Botan. Jahrb. **86,** 402—448 (1967).

BERTSCH, A.: Planta **70,** 46—72 (1966).

BIEBL, R.: Jahrb. wiss. Botan. **86,** 350—386 (1938).

BIEBL, R.: Protoplasmatische Ökologie der Pflanzen. Wasser und Temperatur, Protoplasmatologia, Bd. XII/1. Wien: Springer 1962.

BIEBL, R., MAIER, R.: Österr. botan. Z. **117,** 176—194 (1969).

BLACK, C. C.: In: Advances in ecological research, vol. 7 (ed. J. B. Cragg), pp. 87—114. London-New York: Academic Press 1971.

BLINKS, L. R.: Manual of phycology. Waltham (Mass.): Chronica Botanica Comp. 1951.

BLÜTHGEN, J.: Allgemeine Klimageographie. Berlin: W. de Gruyter 1964.

BÖHNING, R. H., BURNSIDE, C.: Am. J. Botany **43,** 557—561 (1956).

BOLIN, B.: Sci. American **223** (3), 124—132 (1970).

BORNKAMM, R.: Flora (Jena) **146,** 23—67 (1958).

BOYER, J. S.: Planta **117,** 187—207 (1974).

BOYER, Y.: Vie et Milieu, Ser. C.: Biologie terrestre **19,** 331—344 (1968).

BOYSEN-JENSEN, P.: Die Stoffproduktion der Pflanzen. Jena: G. Fischer 1932.

BROCK, Th. D.: Science **158**, 1012–1019 (1967).

BUCH, K.: In: Handbuch der Pflanzenphysiologie, Bd. V/1 (ed. W. Ruhland), pp. 12–23. Berlin-Göttingen-Heidelberg: Springer 1960.

BÜNNING, E.: Entwicklungs- und Bewegungsphysiologie der Pflanze, 3. Aufl. Berlin-Göttingen-Heidelberg: Springer 1953.

CERNUSCA, A.: In: Ökosystemforschung (ed. H. Ellenberg, pp. 195–201, Berlin-Göttingen-Heidelberg: Springer 1973.

CERNUSCA, A.: Umschau **75** (8), 242–245 (1975).

CHRISTOPHERSEN, J.: In: Temperatur und Leben (ed. H. Precht, J. Christophersen, H. Hensel). Berlin-Göttingen-Heidelberg: Springer 1955.

CINTRON, G.: In: A tropical rain forest (ed. H. T. Odum, R. F. Pigeon), pp. H 133–136. Oak Ridge: USAEC 1970.

CLARK, J.: Photosynthesis and respiration in white spruce and balsam fir. State Univ. Coll. For. Syracuse (N.Y.) 1961.

CLODIUS, S.: Biologieunterr. **6**, 101–110 (1970).

COWAN, I. R.: J. Appl. Ecol. **2**, 221–239 (1965).

DÄSSLER, H. G., RANFT, H.: Flora (Jena), Abt. B **158**, 454–461 (1969).

DÄSSLER, H. G., RANFT, H., REHN, K. H.: Flora (Jena) **161**, 289–302 (1972).

DEEVEY, E. S.: Sci. American **223** (3), 148–158 (1970).

DELWICHE, C. C.: Sci. American **223** (3) 136–146 (1970).

DE WIT, C, T.: Verslag Landbouwk. Onderzoek. **64.6** 5–88 (1958).

DE WIT, C. T., BROUWER, R., PENNING DE VRIES, F. W. T.: Proc. IBP/PP Techn. Meeting Třeboň, pp. 47–70. Centre Agr. Publ. Doc., Wageningen 1970.

DONALD, C. M.: Symp. Soc. for Exp. Biol. **15**, 282–313 (1961).

DÖRING, B.: Botan. Z. **28**, 305–383 (1935).

DUVIGNEAUD, P. (ed.):Ecosystèmes et biosphère. Brüssel: Min. Educ. Nat. Cult. 1967.

DUVIGNEAUD, P., DENAEYER-DE SMET, S.: In: Analysis of temperate forest ecosystems. Ecological studies, vol. 1 (ed. D. E. Reichle), pp. 199–225. Berlin-Heidelberg-New York: Springer 1970.

DUVIGNEAUD, P., DENAEYER-DE SMET, S., AMBROES, P., TIMPERMAN, J., MARBAISE, J. L.: Bull. soc. roy botan. Belg. **102**, 317–327 et 339–354 (1969).

EBERHARDT, F.: Planta **45**, 57–68 (1955).

ECKARDT, F. E., HEIM, G., METHY, M., SAUGIER, B., SAUVEZON, R.: Oecol. Plant. **6**, 51–100 (1971).

ELLENBERG, H.: Mitt. floristisch-soziol. Arbeitsgem. Niedersachsen **5**, 3–135 (1939).

ELLENBERG, H.: Handbuch der Pflanzenphysiologie, Bd. IV (ed. W. Ruhland), pp. 638–708. Berlin-Göttingen-Heidelberg: Springer 1958.

ELLENBERG, H.: Vegetation Mitteleuropas mit den Alpen. Stuttgart: E. Ulmer 1963.

ELLENBERG, H.: Ber. deut. botan. Ges. **77**, 82–92 (1964).

EL-SHARKAWY, M. A., HESKETH, J. D.: Crop. Sci. **4**, 514–518 (1964).

FINCK, A.: Pflanzenernährung in Stichworten. Kiel: F. Hirt 1969.

FINDENEGG, J.: Schweiz. Z. Hydrobiol. **29**, 125–144 (1967).

FINDENEGG, J.: Z. Wasser- und Abwasserforsch. **2**, 139–144 (1969).

FORTESCUE, J. A. C., MARTEN, G. G.: Analysis of temperate forest ecosystems. Ecological studies, vol. 1 (ed. D. E. Reichle), pp. 173–198. Berlin-Heidelberg-New York: Springer 1970.

FRENCH, C. S., YOUNG, V. M. K.: Radiation Biol. **3**, 343–391 (1956).

FREY-WYSSLING, A.: Stoffwechsel der Pflanzen, 2. Aufl. Zürich: Büchergilde Gutenberg 1949.

Fritts, H. C., Blasing, T. J., Hayden, B. P., Kutzbach, J. E.: J. Appl. Meteorol. **10**, 845—864 (1971).

Gaastra, P.: Mededeel. Landbouwhogeschool Wageningen **59**, 1—68 (1959).

Gabrielsen, E. K.: Physiol. Plantarum **1**, 5—37 (1948).

Garber, K.: Luftverunreinigung und ihre Wirkungen. Berlin: Gebr. Borntraeger 1967.

Gates, M.: Ecology **46**, 1—14 (1965).

Geiger, R.: Das Klima der bodennahen Luftschicht, 4. Aufl. Braunschweig: F. Vieweg & Sohn 1961.

Geiger, R.: Die Atmosphäre der Erde. Karte 1: Jährliche Sonnenstrahlung; Karte 5: Jährlicher Niederschlag; Karte 6: Jährliche wirkliche Verdunstung. Darmstadt: Perthes 1965.

Gessner, F.: Hydrobotanik, Bd. I: Energiehaushalt. Berlin: VEB Deutscher Verlag d. Wiss. 1955.

Gessner, F.: In: Handbuch der Pflanzenphysiologie, Bd. IV (ed. W. Ruhland), pp. 179—232. Berlin-Göttingen-Heidelberg: Springer 1958.

Gessner, F.: Hydrobotanik, Bd. II: Stoffhaushalt. Berlin: VEB Deutscher Verlag d. Wiss. 1959.

Gloser, J.: Photosynthetica **1**, 171—178 (1967).

Goryshina, T. K.: Oecol. Plant. **7**, 241—258 (1972).

Gradmann, H.: Jahrb. wiss. Botan. **69**, 1—100 (1928).

Grieve, B. J., Hellmuth, E. O.: Proc. Ecol. Soc. Austral. **3**, 46—54 (1968).

Grin, A. M.: Umschau **72** (17), 551—554 (1972).

Harrasser, J.: Diss. Innsbruck 1969.

Härtel, O.: Protoplasma **34**, 489—514 (1940).

Hellmuth, E. O.: J. Ecol. **59**, 225—259 (1971).

Hesketh, J., Baker, D.: Crop. Sci. **7**, 285—293 (1967).

Hiroi, T., Monsi, M.: J. Fac. Sci., Tokyo **9**, 241—285 (1966).

Hoffmann, G.: Flora (Jena) **161**, 303—319 (1972).

Höfler, K.: Ber. deut. botan. Ges. **38**, 288—298 (1920).

Höfler, K.: Ber. deut. botan. Ges. **60**, (94)—(10) (1942).

Höfler, K.: Ber. deut. botan. Ges. **63**, 3—10 (1950).

Höfler, K., Migsch, H., Rottenburg, W.: Forschungsdienst **12**, 50—61 (1941).

Horak, O., Kinzel, H.: Österr. botan. Z. **119**, 475—495 (1971).

Huber, B.: Die Saftströme der Pflanzen. Berlin-Göttingen-Heidelberg: Springer 1956.

Iljin, W. S.: Jahrb. wiss. Botan. **66**, 947—964 (1927).

Iljin, W. S.: Protoplasma **10**, 379—414 (1930).

Iwaki, H., Midorikawa, B.: In: Methods of productivity studies in root system and rhizosphere organisms (eds. M. S. Ghilarov, V. A. Kovda, L. N. Novichkova-Ivanova, L. E. Rodin, V. M. Sveshnikova), pp. 72—78. Leningrad: Nauka 1968.

Kainmüller, Ch.: Diss. Innsbruck 1974.

Kairiukštis, L. A.: In: Svetovoi režim fotosintez i produktivnost lesa (ed. Ju. L. Celniker), pp. 151—166. Moskau: Nauka 1967.

Kalle, K.: In: Handbuch der Pflanzenphysiologie, Bd. IV (ed. W. Ruhland), pp. 170—178. Berlin-Göttingen-Heidelberg: Springer 1958.

Kallio, P., Heinonen, S.: Rep. Kevo Subarctic Res. Sta. **8**, 63—72 (1971).

Kausch, W.: Planta **45**, 217—265 (1955).

Keller, R.: Umschau **71** (3), 73—78 (1971).

Keller, Th.: Allgem. Forst- u. Jagdztg. **142** (4) (1971 a).

Keller, Th.: Ber. Eidg. Anst. f. forstl. Versuchswesen, Birmensdorf Nr. **67**, (1971b).

Kiendl, J.: Ber. deut. botan. Ges. **66**, 246—263 (1953).

KINZEL, H.: Ber. deut. botan. Ges. **82,** 143—158 (1969).

KIRA, T., OWAGA, H., YODA, K., OGINO, K.: Botan. Mag. (Tokyo) **77,** 428—429 (1964).

KIRA, T., SHIDEI, T.: Japan. J. Ecol. **17,** 70—87 (1967).

KIRA, T., SHINOZAKI, K., HOZUMI, K.: Plant and Cell Physiol. **10,** 129—142 (1969).

KLUGE, M.: Ber. deut. botan. Ges. **84,** 417—424 (1971).

KNAPP, G., KNAPP, R.: Landwirtsch. Jahrb. Bayern **29,** 239—256 (1952).

KNAPP, R.: Experimentelle Soziologie und gegenseitige Beeinflussung der Pflanzen. Stuttgart: E. Ulmer 1967.

KOHLER, A.: Ber. deut. botan Ges. **84,** 713—720 (1971).

KOL, E.: Kryobiologie, I. Kryovegetation. Stuttgart: Schweizerbart 1968.

KRAMER, P. J.: Plant and Soil Water Relationship. New York-Toronto-London: McGraw-Hill 1949.

KRAMER, P. J.: In: The physiology of forest trees, (ed. V. Thimann). New York: Ronald Press Co. 1958.

KRASAVCEV, O. A.: Deut. Akad. Landwirtschaftswiss. Berlin, Tagungsber. **100,** 23—34 (1968).

KREEB, K.: Angew. Botan. **39,** 1—15 (1965).

KREEB, K.: Ökophysiologie der Pflanzen. Jena: VEB G. Fischer 1974.

KRÜSSMANN, G.: Taschenbuch der Gehölzverwendung, 2. Aufl. Berlin-Hamburg: P. Parey 1970.

KYRIAKOPOULOS, E., LARCHER, W.: Z. Pflanzenphysiol. **75,** in press (1975).

LAATSCH, W.: Dynamik der mitteleuropäischen Mineralböden. Dresden: Steinkopff 1954.

LANGE, O. L.: Flora (Jena) **140,** 39—97 (1953).

LANGE, O. L.: Flora (Jena) **147,** 595—651 (1959).

LANGE, O. L.: Ber. deut. botan. Ges. **75,** 351—352 (1962).

LANGE, O. L.: Planta **64,** 1—19 (1965).

LANGE, O. L.: Flora (Jena), Abt. B **158,** 324—359 (1969).

LANGE, O. L., SCHULZE, E. D., KOCH, W.: Flora (Jena) **159,** 38—62 (1970).

LARCHER, W.: Planta **44,** 607—638 (1954).

LARCHER, W.: Bull. Research Council Israel, Sect. D **8,** 213—224 (1960).

LARCHER, W.: Planta **56,** 575—606 (1961).

LARCHER, W.: Planta **60,** 1—18 (1963a).

LARCHER, W.: Protoplasma **57,** 569—587 (1963b).

LARCHER, W.: Klima und Pflanzenleben in Arco. Kulturelle Veröffentlichung. d. Kurverwaltung Arco, II. Temi, Trient 1964.

LARCHER, W.: Photosynthetica **3,** 167—198 (1969 a).

LARCHER, W.: Ber. deut. botan. Ges. **82,** 71—80 (1969 b).

LARCHER. W.: Oecol. Plant. **5,** 267—286 (1970).

LARCHER. W.: Ber. deut. botan. Ges. **85,** 315—327 (1972).

LARCHER. W.: In: Temperature and life, 2. eds. H. PRECHT, J. CHRISTOPHERSEN, H. HENSEL, W. LARCHER). Berlin-Heidelberg-New York: Springer 1973.

LARCHER. W., CERNUSCA, A., SCHMIDT, L.: In: Ökosystemforschung (ed. H. ELLENBERG), pp. 175—194. Berlin-Heidelberg-New York: Springer 1973.

LARCHER. W., MAIR, B.: Oecol. Plant. **3,** 255—270 (1968).

LARCHER. W., MAIR, B.: Oecol. Plant. **4,** 347—376 (1969).

LEVITT, J.: The hardiness of plants. New York: Academic Press 1956.

LEVITT, J.: Frost, drought, and heat resistance. Protoplasmatologia, vol. VIII/6. Wien: Springer 1958.

LIETH, H.: In: Analysis of temperate forest ecosystems. Ecol. studies, vol. 1 (ed. D. E. Reichle), pp. 29—46. Berlin-Heidelberg-New York: Springer 1970.

LIETH, H.: Angew. Botan. **46**, 1–37 (1972).

LORENZEN, H.: Physiologische Morphologie der Pflanzen (UTB 65). Stuttgart: E. Ulmer 1972.

LUDLOW, M. M., WILSON, G. L.: Australian J. Biol. Sci. **24**, 449–470 (1971a).

LUDLOW, M. M., WILSON, G. L.: Australian J. Biol. Sci. **24**, 1065–1075 (1971b).

LUDLOW, M. M., WILSON, G. L.: Australian J. Biol. Sci. **24**, 1077–1087 (1971c).

MALKINA, I. S., CELNIKER, JU. L., JAKŠINA, A. M.: Fotosintez i dychanie podrosta. Moskau: Nauka 1970.

MARCELLO, A.: Nuovo giorn. botan. ital., N.S. **66**, 929–1034 (1959).

MAR-MÖLLER, C., MÜLLER, D., NIELSEN, J.: Det. Forstl. Forsøgsv. i Danmark **21**, 327–335 (1954).

MEIDNER, H., MANSFIELD, T. A.: Physiology of stomata. London: McGraw-Hill 1968.

MENGEL, K.: Ernährung und Stoffwechsel der Pflanze, 3. Aufl. Jena: VEB Fischer 1968.

MESSERI, A.: Nuovo giorn. botan. ital., N. S. **58**, 535–549 (1951).

MILLER, R., RÜSCH, J.: Forstwiss. Centr. **79**, 42–62 (1960).

MITSCHERLICH, G.: Wald, Wachstum und Umwelt. 2. Bd.: Waldklima und Wasserhaushalt. Frankfurt (Main): Sauerländer 1971.

MONSI, M., SAEKI, T.: Japan. J. Botany **14**, 22–52 (1953).

MOONEY, H. A., WEST, M.: Am. J. Botany **51**, 825–827 (1964).

MONTEITH, J. L.: Ann. Botany (London) **29**, 17–37 (1965).

MÜLLER, D.: In: Handbuch der Pflanzenphysiologie, Bd. XII/2 (ed. W. Ruhland), pp. 934–948. Berlin-Göttingen-Heidelberg: Springer 1960.

MÜLLER, E., LOEFFLER, W.: Mykologie, 2. Aufl. Stuttgart: G. Thieme 1971.

MÜLLER-STOLL, W. R.: Z. Botan. **29**, 161–253 (1935).

NEGISI, K.: Bull. Tokyo Univ. Forests **62**, 1–115 (1966).

NEUWIRTH, G.: Biol. Zentr. **78**, 560–584 (1959).

ODUM, E. P.: Fundamentals of ecology. 3rd ed. Philadelphia-London: Saunders 1971.

OPPENHEIMER, H. R.: Ber. deut. botan. Ges. **50 a**, 185–243 (1932).

PARKER, J.: NE. Forest Exp. Sta. Upper Darby, Pa., US-Forest Serv. Res. Paper NE-94, 1968.

PISEK, A.: Gartenbauwissenschaft **23**, 54–74 (1958).

PISEK, A., BERGER, E.: Planta **28**, 124–155 (1938).

PISEK, A., CARTELLIERI, E.: Jahrb. wiss. Botan. **75**, 195–251 (1931).

PISEK, A., CARTELLIERI, E.: Jahrb. wiss. Botan. **75**, 643–678 (1932).

PISEK, A., CARTELLIERI, E.: Jahrb. wiss. Botan. **79**, 131–190 (1933).

PISEK, A., CARTELLIERI, E.: Jahrb. wiss. Botan. **90**, 256–291 (1941).

PISEK, A., KEMNITZER, R.: Flora (Jena) Abt. B **157**, 314–326 (1968).

PISEK, A., KNAPP, H., DITTERSTORFER, J.: Flora (Jena) **159**, 459–479 (1970).

PISEK, A., SCHIESSL, R.: Ber. naturwiss.-med. Ver. Innsbruck **47**, 33–52 (1947).

PISEK, A., TRANQUILLINI, W.: Physiol. Plantarum **4**, 1–27 (1951).

PISEK, A., WINKLER, E.: Planta **42**, 253–278 (1953).

PISEK, A., WINKLER, E.: Protoplasma **46**, 597–611 (1956).

PISEK, A., WINKLER, E.: Planta **51**, 518–543 (1958).

PISEK, A., WINKLER, E.: Planta **53**, 532–550 (1959).

POLSTER, H.: Die physiologischen Grundlagen der Stofferzeugung im Walde. München: Bayer. Landwirtschaftsverlag 1950.

POLSTER, H.: In: Gehölzphysiologie (eds. H. Lyr, H. Polster, H. J. Fiedler). Jena: VEB G. Fischer 1967.

POLSTER, H., FUCHS, S.: Arch. Forstwesen **12**, 1011–1024 (1963).

POLSTER, H., NEUWIRTH, G.: Arch. Forstwesen **7**, 749–785 (1958).

RAEUBER, A., MEINL, G., ENGEL, K. H.: Wiss. Z. Karl-Marx-Univ. Leipzig, Math.-naturw. Reihe 17, 295–301 (1968).

RANFT, H., DÄSSLER, H. G.: Flora (Jena) 159, 573–588 (1970).

RAPP, M.: Oecol. Plant. 4, 377–410 (1969).

RAPP, M.: Cycle de la matière organique et des éléments minèraux dans quelques écosystèmes méditerrannèens, vol. 40, pp. 19–184. Paris: Édit. C.N.R.S. 1971.

RASCHKE, K.: Planta 48, 200–239 (1956).

REHDER, H.: Diss. Bot. 6, Lehre: J. Cramer 1970.

RETTER, W.: Diss. Innsbruck 1965.

RICHTER, H., HALBWACHS, G., HOLZNER, W.: Flora (Jena) 161, 401–420 (1972).

ROOK, D. A.: New Zealand J. Bot. 7, 43–55 (1969).

ROUSCHAL, E.: Österr. botan. Z. 87, 42–50 (1938).

RUTTER, A. J.: In: Water deficit and plant growth, vol. II (ed. T. T. Kozlowski), pp. 23–84. New York-London: Academic Press 1968.

RYCHNOVSKÁ, M.: Preslia 37, 42–52 (1965).

RYCHNOVSKÁ, M., KVĚT, J., GLOSER, J., JAKRLOVÁ, J.: Acta Sci. Nat. Acad. Sci. Bohemoslovaceae Brno 6 (5), (1972).

SALISBURY, E. I.: J. Ecol. 4, 83–117 (1916).

SAUBERER, F., HÄRTEL, O.: Pflanze und Strahlung. Leipzig: Akad. Verlagsgesellschaft 1959.

SCHEFFER, F.: In: Handbuch der Pflanzenphysiologie, Bd. VIII (ed. W. Ruhland), pp. 179–200. Berlin-Göttingen-Heidelberg: Springer 1958.

SCHNELLE, F.: Pflanzenphänologie. Leipzig: Verlagsgesellschaft 1955.

SCHNOCK, G.: Oecol. Plant. 7, 205–226 (1972).

SCHÖLM, H. E.: Protoplasma 65, 97–118 (1968).

SCHOLANDER, P. F., HAMMEL, H. T., BRADSTREET, E. D., HEMMINGSEN, E. A.: Science 148, 339–346 (1965).

SCHROEDER, D.: Bodenkunde in Stichworten. Kiel: F. Hirt 1969.

SCHULZE, E. D.: Flora (Jena) 159, 177–232 (1970).

SCHULZE, E. D., LANGE, O. L., KOCH, W.: Oecologia 9, 317–340 (1972).

SCHULZE, R.: Strahlenklima der Erde. Darmstadt: Steinkopff 1970.

ŠESTÁK, Z., ČATSKÝ, J., JARVIS, P. G.: Plant photosynthetic production. Manual of methods. Den Haag: W. Junk 1971.

SIEGELMAN, H. W., BUTLER, W. L.: Ann. Rev. Plant Physiol. 16, 383–392 (1965).

SLATYER, R. O.: Plant-water relationships. London-New York: Academic Press 1967.

SLAVIK, B. (ed.): Methods in studying plant water relations. Ecological Studies vol. 9. Berlin-Heidelberg-New York: Springer 1974.

SMITH, P. F.: Ann. Rev. Plant Physiol. 13, 81–108 (1962).

STÅLFELT, M. G.: Planta 27, 30–60 (1937).

STANHILL, G.: In: Analysis of temperate forest ecosystems. Ecological studies, vol. 1 (ed. D. E. Reichle), pp. 247–256. Berlin-Heidelberg-New York: Springer 1970.

STEUBING, L., DAPPER, H.: Ber. deut. botan. Ges. 77, 71–74 (1964).

STOCKER, O.: Tabulae Biologicae, Bd. V, pp. 510–686. Berlin: W. Junk 1929.

STOCKER, O.: In: Handbuch der Pflanzenphysiologie, Bd. III (ed. W. Ruhland), pp. 436–488. Berlin-Göttingen-Heidelberg: Springer 1956.

STOCKER, O.: Flora (Jena) 159, 539–572 (1970).

STOCKER, O.: Flora (Jena) 161, 46–110 (1972).

STOY, V.: Physiol. Plantarum 4, 1–125 (1965).

STOY, V.: Ber. Arbeitstagung Gumpenstein pp. 29–49, 1966.

SULLIVAN, C. J., LEVITT, J.: Physiol. Plantarum 12, 299–305 (1959).

TAIT, R. V.: Elements of marine ecology. London: Butterworth & Co. 1968.

TALLING, J. F.: Proc. IBP/PP. Techn. Meeting Trebon 1969, pp. 431–445, Centre Agr. Publ. Doc. Wageningen 1970.

THOMPSON, P. A.: Nature 217, 1156–1157 (1968).

TILL, O.: Flora (Jena) 143, 499–542 (1956).

TOTSUKA, T.: Fac. Sci., Univ. Tokyo 8, 341–375 (1963).

TRANQUILLINI, W.: Planta 54, 130–151 (1959).

TRANQUILLINI, W., SCHÜTZ, W.: Centr. ges. Forstw. 87, 42–60 (1970).

TRANQUILLINI, W., UNTERHOLZNER, R.: Centr. ges. Forstw. 85, 97–110 (1968).

TREHARNE, K. J., COOPER, J. P.: J. Exptl. Botany 20, 170–175 (1969).

TREHARNE, K. J., EAGLES, C. F.: Photosynthetica 4, 107–117 (1970).

TRESHOW, M.: Environment and plant response. New York-London-Toronto: McGraw Hill 1970.

TROLL, C.: Studium Generale 8, 713–733 (1955).

TRUOG, E.: Soil Sci. Soc. Am. Proc. 11, 305–308 (1947).

ULLRICH, H., MÄDE, A.: Planta 31, 251–263 (1940).

URSPRUNG, A., BLUM, G.: Ber. deut. botan. Ges. 36, 599–618 (1918).

VARESCHI, V.: Planta 40, 1–35 (1951).

WALTER, H.: Die Hydratur der Pflanze und ihre physiologisch-ökologische Bedeutung. Jena: G. Fischer 1931.

WALTER, H.: Einführung in die Phytologie, Bd. III/1. Standortslehre, 2. Aufl. Stuttgart: E. Ulmer 1960.

WALTER, H.: Z. Pflanzenphysiol. 56, 170–185 (1967).

WALTER, H.: Die Vegetation der Erde in ökophysiologischer Betrachtung, Bd. II: Die gemäßigten und arktischen Zonen. Jena: VEB G. Fischer 1968.

WALTER, H.: Vegetationszonen und Klima (UTB 14). Stuttgart: E. Ulmer 1970.

WALTER, H., LIETH, H.: Klimadiagramm-Weltatlas. Jena: VEB G. Fischer 1967.

WAREING, P. F., PHILLIPS, I. D. J.: The control of growth and differentiation in plants. Oxford: Pergamon Press 1970.

WENT, F. W.: The experimental control of plant growth. Waltham (Mass.): Chronica Botanica Comp. 1957.

WHITTAKER, R. H.: Communities and ecosystems. London: Collier-Macmillan Ltd 1960.

WILSON, CH. CH.: Plant Physiol. 23, 5–35 (1948).

YOCUM, C. S., ALLEN, L. H., LEMON, E. R.: Agron. J. 56. 249–253 (1964).

YODA, K.: Nature and Life in Southeast Asia 5, 83–148 (1967).

YODA, K., KIRA, T.: Nature and Life in Southeast Asia 6, 83–110 (1969).

ZELNIKER, J. L.: Deut. Akad. Landwirtschaftswiss. Berlin, Tagungsber. 100, 131–140 (1968).

Subject Index

Page numbers in bold face refer to the most important catch words, page numbers with an asterisk refer to figures, tables and formulas.

245

H. Walter
Vegetation of the Earth

in Relation to Climate and the Eco-Physiological
Conditions. Translator: J. Wieser
79 figures. XV, 237 pages. 1973 (Heidelberg Science
Library, Vol. 15)
ISBN 3-540-90046-2 DM 16,–
ISBN 0-387-90046-2 (North America) $5.90

Distribution rights for U.K., Commonwealth, and the
Traditional British Market: English Universities
Press Ltd., London

Prices are subject to change without notice

Contents: Evergreen Tropical Rain Forest Zone Vegetation
of the Tropical Summer-Rain Zone. Subtropical Semidesert
and Desert Zones. Sclerophyllous Vegetation of the
Regions with Winter Rains. The Warm-Temperate Vegetational Zone. Nemoral Zone or Deciduous Forest Zone
in Temperate Climates. The Arid Vegetational Regions
of the Temperate Climatic Zone. The Boreal Coniferous
Forest Zone. The Arctic Tundra Zone. The Alpine
Vegetation of Mountainous Regions.

Plant ecology is basic to general, animal, systems, paleo-,
and human ecology. Plants are the primary producers; they
dominate the flow and cycling of energy, water, and
mineral nutrients within ecosystems.

The structure of the vegetation which plants form
determines much of the character of the landscapes in
which other organisms, including men and women, live
and prosper. If we know why plants grow where they do,
we know a good deal about why organisms other than
plants live where they do.
This volume concentrates on descriptions of kinds of
zonal vegetation into which eco-physiological data must
fit. It presents this information in a consistent frame of
reference and suggests where more factual data are
needed.
An excellent introduction to plant ecophysiology, it
analyzes past research in the field and makes clear how
much more investigation must be undertaken to describe
plant and vegetation structure and function, their
correlations, and their relationships to environment.

Springer-Verlag
Berlin
Heidelberg
New York

D. Hess
Plant
Physiology

Molecular, Biochemical, and Physiological Fundamentals of Metabolism and Development

Springer Study Edition
248 figures. XV, 333 pages. 1975
ISBN 3-540-06643-8 DM 36,30
ISBN 0-387-06643-8
(North America) $14.80
Prices are subject to change without notice
Distribution rights for India: UBS, New Delhi

Contents
Control of Character Formation by Nucleic Acids. –
Photosynthesis. – Carbohydrates. – Biological Oxi-
dation. – Fats. – Terpenoids. – Phenols. – Amino
Acids. – Alkaloids. – Perphyrins. – Cell Division. –
Differential Gene Activity as Principle of Differen-
tiation. – Regulation. – Polarity and Unequal Cell
Division as Fundamentals of Differentiation. –
Cell Elongation. – The Formation of Seeds and
Fruits. – Germination. – The Vascular System. –
Flower Formation.

Springer-Verlag
Berlin
Heidelberg
New York
An elementary introduction to the metabolic and
developmental physiology of higher plants from
the point of view of molecular biology. The out-
standing feature of the text is that equal weight is
given to metabolism and development.